SCIENCE NONFICTION

Behind the Scenes in University Research

Darren Lipomi

ISBN: 979-8-9957969-0-9 (hardcover)
ISBN: 979-8-9957969-1-6 (paperback)
ISBN: 979-8-9957969-2-3 (ebook)

To Dad

Prologue

"So you're a professor?" the immigration officer said. He held up my passport to see if the photo matched my appearance, minus the tousled hair and dark circles under my eyes.

It was six in the morning, and I'd just stepped off a twelve-hour flight from Singapore to San Francisco. I was headed back home, to San Diego, after a week of talks at universities in three countries.

The officer glanced up from the screen. "What do you teach?"

That question always throws me. There's an easy answer: *chemical engineering*. Then there's the real one. Yes, I teach students about chemistry, engineering, and materials science. But I spend most of my time raising money so they can do experiments that are more likely than not to fail. I teach undergraduates, PhD students, and postdoctoral researchers how to ask good scientific questions, develop strong written arguments, speak in public, and not lose hope when their manuscripts are rejected by journal editors and unkind reviewers.

"Chemical engineering" usually seems to satisfy airport protocol, so that's what I went with, and the officer waved me through.

That exchange, repeated in airports, in dentists' chairs, and at holiday parties around the world, captures a truth even people close to me never quite grasped. My father, for instance, believed until the day he died that my job consisted of walking into a classroom every morning and lecturing. He'd never been inside a lab—not even mine. He knew I was a professor; he just didn't know what that meant.

Most people don't. To the outside world, science is either heroic or hopelessly abstract—a lone genius discovering the cure for something or a bureaucratic enterprise burning money in the name of curiosity. The truth sits somewhere in between. In a typical lab discovery happens amid the sickly-sweet smell of acetone and the deafening hiss of overpressurized tanks of liquid nitrogen (don't worry, it's supposed to do that). Every breakthrough sits on a mountain of discarded pipette tips, rubber gloves, and ideas that didn't work out. Progress comes in microns, not miles, and it all takes money—a lot of it.

The United States spends roughly $60 billion a year—about 1.2 percent of the federal budget—on university research. Most of that money doesn't buy machines; it buys *time*. Time for graduate students and postdoctoral researchers to chase hunches and ask big questions, design experiments to test them, and fix equipment that costs more than their cars.

In other words, to learn how to think like scientists.

What the public rarely sees is that research is mostly a human story. The labor of these bright-eyed, bushy-tailed idealists fuels an entire knowledge economy, even though their stipends barely cover rent in most coastal university cities.

These students and postdoctoral scholars give their twenties, and often their thirties, to a system that trains them brilliantly while exhausting them completely.

Nearly half of all students entering PhD programs don't finish.

The system is far from perfect. Universities use student labor to generate revenue and cachet. Students, for their part, seek certification— a PhD diploma—to convince a future employer to pay them enough to feel secure. Along the way, power dynamics between professors and students can tilt toward exploitation. The traits that make someone good at science (obsession, perfectionism, self-critique) can easily curdle into anxiety or despair.

Yet the same environment produces resilience, creativity, and, occasionally, discoveries that transform entire industries—and lives. The line between breakthrough and burnout is as thin as the glass capillary tube all good chemists learn how to make with a blowtorch.

The system is riddled with contradictions. Science is expensive, yet the median salary for the people who do most of the work is less than $40,000. There is a fair case to be made that the financial burden of research is borne not by the taxpayer, but by the "forever trainee"—the twenty-two-year-old PhD student who becomes a postdoctoral scholar at twenty-nine, and an untenured research scientist at thirty-four. These individuals earn a fraction of those with MBAs and JDs, degrees that took much less time to obtain than PhDs and postdoctoral training.

They make other sacrifices as well, such as forgoing the investment interest early in adulthood that leads to a comfortable retirement. Time in the lab—eighty hours a week or more—combined with financial insecurity means that researchers start families later in life. The stress PhD students experience leads to significant attrition among them. Asymmetric power relationships, sometimes between students and their advisors, can lead to abuse. For all these reasons, it is unsurprising that around half of graduate student researchers seek treatment for anxiety and depression, a rate much higher than the general population.

Still, I believe in the enterprise. Nearly every technology that defines modern life—next-generation semiconductors, the internet, life-saving vaccines and anticancer therapeutics, CRISPR gene editing, solar panels, lithium-ion batteries for electric cars and energy independence, and GLP-1 drugs for diabetes and weight loss—had its basis in university labs. When the system works, it's invisible. You only notice it when it fails.

Or possibly when a politician decides to score points by calling research "woke" or "wasteful." This happened with one of my own grants for developing a device to help throat cancer patients recover their ability to swallow. It was singled out by a Senate committee as Marxist because a tiny fraction of the budget—as required by the rules of the program—was allocated to outreach. Now funding for these purposes has an uncertain future.

The politics, the bureaucracy, the rankings—all of it can make the university feel less like an ivory tower and more like an industrial park.

But I've also seen the other side—for example, a graduate student's first successful experiment after months of failure. Or the quiet triumph of a kid who grew up poor and became an addict, got clean, earned a PhD, and is inventing America's future at a big company (and earning much more money than I do). Those moments remind me why I stay.

The system may be messy, but it remains one of society's best inventions—public money invested in some of its most curious, tenacious people. My six-year-old daughter recently asked me if time travel is possible. In a way, yes. A university research lab, though it runs on a shoestring budget, offers a sneak peek of the future most people won't see for decades.

My own journey through that world has been long and circuitous. It began in a working-class family outside of Rochester, New York. My journey wound through Boston University, Harvard, Stanford, and UC San Diego and eventually carried me back home, to the University of Rochester. Along the way I've been the overworked graduate student, the idealistic postdoc, and the new professor. I have been the careful steward of around $10 million in public funding. I've been a classroom instructor, a laboratory director, and a mentor to students who founded three companies.

I've worked deep in both the defense-academic-industrial complex and the diversity-equity-inclusion milieu and received an award from the White House signed by President Trump, of all people. I served as the associate dean for students at the engineering school with the greatest number of students on the West Coast. I've been persistent and profoundly lucky. At Rochester, I'm the department chair of a storied but newly reborn department of chemical and sustainability engineering, trying to keep the next generation of students and faculty from losing hope as it faces an uncertain financial future. I've seen science at its most inspiring and its most absurd.

This book is about that world—the people who make discoveries happen and what the process does to them. It's about their ambition and anxieties (from arrhythmias to hyperhidrosis to lockjaw, I've had them all). It's also about what I've learned after twenty years inside the machine: that science is as human as any art, that universities are held together less by policy than by decency, and that progress, in the end, depends on ordinary people doing extraordinary amounts of work.

The habits that shaped that life—curiosity, persistence, a tolerance for failure—didn't begin in a lab. They started long before I knew what a grant or a publication was, in a small town in Western New York. There I built solar cars that leaked battery acid and messed with the source code of games that came with Windows 3.1 just to see how they worked. To explain the present, I have to go back to the village of Hilton, where the road to academia began with Lego blocks on the floor, *Star Trek* on TV, and the beeps and whir of a 1980 desktop computer.

Chapter 1

I grew up in an affordable tract development in rural small-town Hilton, New York, just a thirty-minute drive outside of Rochester. Easily confused by out-of-staters as a suburb of New York City, Rochester is nearly six hours from the Big Apple (though only ninety minutes to the Canadian border at Niagara Falls). With over a million people and thus the third-largest metro area in New York State, Rochester nevertheless has a reputation for remoteness. (And snow. It is sometimes called Hothchester by locals to call to mind the ice planet in *The Empire Strikes Back*.) Hilton is remote, even by Rochester standards. In fact, it is not even a town but a village, a term of endless amusement for my wife, who associates villages with hobbits and medieval peasant enclaves. "I married a *villager!*" she likes to say. According to the 2024 census, the population of Hilton is hovering around six thousand.

The fact that Hilton was and is small remains one of its defining characteristics. There's a grocery store, and when I was a teenager, they built a McDonald's. There was a bowling alley (Pleasure Lanes), a few barber shops, and my father's tailor and drycleaner shop (Mars-L Cleaners Tailors). There was also a mom-and-pop video rental place, J&M Video, which had a great selection of 8-bit Nintendo games. J&M lost a lot of business when a Blockbuster popped up in a nearby town, and then J&M closed for good when a runaway pickup slammed into its brick facade—an event that I witnessed with my own eyes while in the back seat of my father's car during a snowstorm.

What it lacked in commercial essentials, Hilton made up for in peacefulness, historic architecture, and natural beauty. Sadly, most of its advantages were lost on me as a child. While summertime trips to the sandy and picturesque Hamlin Beach State Park may have defined my childhood, you couldn't buy anything there. If you needed something that you couldn't get from the handful of stores our little town provided, you needed to go to the town of Greece (about a fifteen-minute drive southeast) or even—*gasp!*—downtown Rochester to get it.

My parents grew up in Rochester proper, my dad in a neighborhood populated by immigrants from Southern Italy. The youngest of six, my father, Mariano (though everyone called him Mars), had five older sisters who, along with their husbands, were my de facto Sicilian American grandparents. My paternal grandparents, who were born in Sicily and entered the US at Ellis Island, both passed away in the late 1960s and early 1970s, well before my time. My father had taken over his own father's tailor shop in downtown Rochester and inherited the house he was born in, possibly quite literally.

My mother grew up in a house with the Genesee River in her backyard. Her father was a French Canadian immigrant, and her mother, the only grandparent I ever knew, was of pre–Civil War Irish descent. It was from my mother that I inherited my intellectualism. A world-class reader, she grabbed at knowledge wherever she could find it. Unlike my father, my mother was a college graduate—she held a degree in visual art from the Rochester Institute of Technology, where she also obtained a master's degree. She and my dad married in 1974 at the chapel of a nearby institution, the University of Rochester. Before my parents moved to Hilton, they had my two sisters, Andrea, in 1978, and Deena, in 1979. At the time, my parents were still living in the inner city. By the late 1970s the area was characterized by frequent break-ins and corner drug deals, so they moved "west."

It sometimes surprised me that my mother, with her university education, introspective tastes, and roots in the city, didn't feel out of place in Hilton, given the simplicity of life there. If she felt closed in by the confines of our small town, she never let on. Once my sisters and I left the house, she indulged her love of books, serving several stints as the director at our local Parma Public Library. She completed her second master's degree—this one in library science—at the University at Buffalo, at age sixty-eight. Now, in her eighties, she still reads over a hundred books a year.

Hilton is famous locally for two things: its Apple Festival and how far the parents of student-athletes in other districts must drive to get to our boondock village. We did produce a few pro athletes, such as Ryan Callahan, hockey player and former captain of the New York Rangers (and a silver-medal Olympian), and, much earlier, Cathy Turner, an Olympic speed skater and two-time gold medalist. As a child I went to a so-called homecoming event for Turner at Hilton Fire Department's event hall. Turner was the most famous person I ever got within earshot of as a child.

When I was born, in 1983, apple trees defined our town—remnants left over from the apple-canning industry of the 1800s and 1900s, now decorative features in homeowners' yards.

The Hilton Central School District was another point of town pride. Attracting caring teachers from around the region, it offered myriad extracurriculars and an above-average number of Advanced Placement courses. Its music program sent many students to top-notch conservatories, and its athletic programs often achieved regional success.

My mother and sisters are all artists, and my sisters gravitated toward artistic partners. While I have no skills in visual art, my hobbies have been literary and musical. Some of my students have come to see artistic influences in my scientific work. "Darren's lab," one of my PhD students wrote, years back, in an anonymous letter obtained to support my promotion, "is more like an art studio than an engineering lab."

As elementary school kids, my sisters and I all participated in a selective program called REACH. Led by a visionary teacher, Gordon Lamont, REACH presaged graduate school in the sciences and engineering: There were several yearlong "builds," in the style of YouTube's Mark Rober or the Discovery Channel's *MythBusters*. The builds would culminate in a demo day at the end of the year, commanding awe from all students. In the fourth grade I got to work on a solar-powered car. More than a mere toy, it could be ridden by an adult, like a single-seat golf cart. During demo day, an overzealous rider smashed the car into a curb. The impact ruptured the battery, which leaked acid all over the parking lot. One of my fondest memories of REACH was when, at age ten, I worked with Mr. Lamont after everyone else had left, to get the hazardous, leaking project car in such a state that it could be loaded safely onto his truck.

In Hilton I was a nerdy kid, interested in Lego and scientific franchises that started with the word *Star*—both *Trek* and *Wars*—in equal measure. I often occupied myself with odd gadgets—wires, LEDs, motors—donated to me by a family friend who worked at Kodak, the region's largest employer. I was drawn to the boundless excitement of outer space, and to the infinite permutations of the Legosphere and rudimentary computer programming. You could build anything you wanted. The world I inhabited was full of possibility, whereas the world inhabited by my peers—sports and action figures—felt narrow by comparison.

Still, the impulse to fit in is powerful at that age, and it was a game I couldn't win. My interests made me an outcast, and becoming an outcast was a vicious cycle. I was always the one who was unlike the other boys. When I was as young as five, I scowled when I was given G.I. Joes as gifts (the same way my daughter now recoils when given dolls or "princess stuff"). I bristled at the aggression, the loudness, and the crew cuts of the other boys. Some of the nerdy kids found their tribe in theater. For me, it was band. I was proud to be a band geek (my instrument was trombone), and all my friends were in band. I wore the ridicule of the jocks when the band marched by them to perform the halftime show on Friday nights as a badge of honor.

At any rate, I did not like doing what was, in the 1980s and 1990s, considered "boy stuff," and sports mostly fell into this category. Born with profound astigmatism and outfitted with huge gold-framed glasses, I had the depth perception of a goat and wouldn't have been able to catch a football if my life depended on it. I didn't even like to go outside when the neighborhood boys were playing football in the yard across the street, afraid I'd be called *fag*, a word that was ruthlessly common back then. So I stayed inside to read, play Nintendo, and deconstruct and build games on QBASIC, the programming environment that came with IBM-compatible PCs in the early 1990s.

In addition to the shame of my social and athletic ineptitude, I harbored shame about my older father, who walked with a shuffle and needed to heave himself up from an armchair just to stand. He was fifty-two when I was born. I remember that when I was little, my dad was intensely funny, making faces and indulging in slapstick. Once, during a summer water gun fight, he emerged from the house wearing a hooded poncho and brandishing a Super Soaker, impervious to our direct assaults, like the Terminator. We gleefully blasted his vulnerable pant legs below the knee. He didn't care that his work pants were now soaked.

By the time I was ten, however, I started to become aware of how old he was compared to the other boys' dads. He was sixty-two, and he wasn't a vibrant sixty-two either. He felt, to me, almost elderly, more like a grandparent than a parent. That I saw him this way feels sad—that I was unable to extend him grace—but it was an unvarnished truth: I felt shame for him then, and I feel a different shame now, shame for how I felt *about* him. He just looked old—so much older than William Shatner, who played Captain Kirk, my hero, on the screen and who was exactly the same age as he was. I remember, when we played catch, running over to my dad to pick up the ball when he dropped it (or when I threw it past him), so he didn't have to bend over to pick it up. He did this so slowly that it was agonizing to watch. Having friends over at my house was always awkward because it meant introducing people to my father, who I envisioned would embarrass me by saying something old.

One of his favorite expressions—one that always put me on edge—had to do with two-dollar bills, already out of circulation in my youth and now an archaic form of currency. "Those are *who-ur* notes," my father said of the bills. I'm not sure what was more humiliating: the use of the word *whore*, the antiquated pronunciation of it as *who-ur*, or the description of money as *notes*, a word so old-timey that it felt like it belonged in a black-and-white movie.

Notwithstanding the *who-ur* notes, most of this anxiety regarding my dad's age lived squarely with me. Once my friends got over the fact that he was my father and not my grandfather, they enjoyed the interest he took in their hobbies, and he lingered in their memory. While he appeared to be an epic schmoozer at gatherings of friends and family— and at the Greek diner and the Old Country Buffet—his interest in others was genuine. His interactions with nearly everyone were memorable, "legendary" even, as expressed to me by my cousin Carol's sons, Michael and Anthony, at his funeral. When he died, in 2021, twenty years after I left Hilton, I received calls of consolation from high school friends, filling me with pride but also fueling my guilt for how I had regarded him back then.

Growing up I didn't know a lot about my family's financial situation, or what I knew was sparse—for instance, that there were limits to what we could have, though I had no broader sense of how dire things were. Sometimes when we asked for simple pleasures, we were denied, but not for conventional reasons. If our parents said no to a thirty-five-cent candy bar, it wasn't because they objected to sugar or junk food; it was because they genuinely couldn't afford the expenditure. Only in retrospect do I understand how little they must have lived on, and their thin margin of security. According to tax returns filed in the early 1990s, our family's adjusted gross income one year was just $2,100. My father did run his own business, so it's impossible to know exactly what he wrote off; even with write-offs and exemptions because of three kids, it's clear that our income was low.

My father died in 2021 following a ten-year battle with dementia and heart disease. In addition to running his dry-cleaner-tailor shop, he sold real estate on the weekends, just to make ends meet. We never took vacations, as there was neither money nor time for luxuries outside of work. He didn't know anything about academia, hadn't read Orwell or Sagan, and hadn't been taught much about science, though he had a natural talent for mental arithmetic. Neither he nor any of his sisters had gone to college. I often had to explain my world to him. Even when I received my acceptance letter to Harvard's PhD program and I called home, he was incredulous.

"When are you going to get the *real* letter?" he wanted to know.

"This is the real letter," I said.

For most of my childhood the only television we had was one my mother had won, being the eleventh caller in a contest held by the local news radio station, 1180 WHAM. We never had cable until we acquired it by accident in 1996, when we signed up for a free monthlong promotion. After we declined—too expensive—they never sent a technician to cancel the service.

As the youngest of three siblings, I had a different experience growing up than my sisters did. We may not have had much money, but I felt nurtured. My oldest sister, Andrea, didn't. At fourteen she met Ronnie, a tall, bespectacled, and intense sixteen-year-old who would shape our family for the next three decades. As renegade adolescents, Andrea and Ronnie were combustible. Andrea experienced the litany of teenage crises: choices and consequences too heavy for her age, expulsion from high school at sixteen, and multiple attempts to run away; some thwarted by the police, others successful. She drifted through her teenage years and turned her anger inward and outward—mostly in the form of shouting matches with my father. When she finally moved in with Ronnie at eighteen, it was to the very same rough neighborhood in Rochester from which my parents had escaped fifteen years earlier. From when I was thirteen to twenty-one, she and Ronnie lived on the margins, mostly invisible to me and completely invisible to our mother and father.

Andrea reappeared in crisis. When prostate cancer surgery and a broken kneecap put my father and mother out of commission in 2004, she showed up at their door with laundry detergent and quiet resolve, as if to make amends. Her story evened out after that. She built a life and business in Las Vegas—but tragedy returned. Ronnie's brother Frank, a generous and hilarious guy loved by both his brother and my sister, died by suicide in 2014. Ronnie unraveled over the ensuing years, and a year after he and my sister divorced, he ended his own life. Ronnie's death in 2021 ended a long season of volatility that began when I was ten and lasted well into adulthood. It was also the first event in a tumultuous year that would lead to my return to Rochester in 2024 and my sister's return in 2025.

It was the windfall of selling my overpriced house in San Diego that allowed me to help my sister relocate. Watching her reestablish herself as a student in social work and spend time with our mother and her two nieces, I feel a profound sense of peace and recognition of our family's resilience. In many ways, if the instability of this aspect of my home life offered one kind of education, science became the other: a counterweight, a framework for making sense of the world.

Chapter 2

My childhood was also a time of discovery and exploration. When I was five years old, Santa splurged and brought me a gift that would change my life: a Fisher-Price PXL2000, a "toy" video camera that could capture eleven minutes of black-and-white video on each side of an audio cassette. At a time when I couldn't even lift the weight of a grown-up's full-size VHS camcorder, the one-hundred-dollar PXL2000 was a lightweight marvel of point-and-shoot simplicity. While my filmmaking ambitions rarely went beyond making my stuffed animals act out scenes from *Star Wars*, these videos presaged my YouTube channel.

During these years I also found science. I found music. Around 1990, when I was seven, my parents received a computer from a family friend who worked at Kodak. In Rochester there were three major industrial employers—Kodak, Xerox, and Bausch & Lomb, Rochester's "Big Three." As Detroit's titans led the world in automotives, Rochester's led the world in imaging. Kodak had a technical workforce in the tens of thousands. Scientists and engineers there were constantly innovating. While working at Kodak in the mid-1980s, my now-colleague at the University of Rochester, Professor Emeritus Ching W. Tang, invented the organic light-emitting device (OLED), the technology used in iPhone screens and the best televisions. Such was the largesse at Kodak that they often dumped outdated equipment or at least encouraged their personnel to abscond with it.

My father's friend and jack-of-all-trades Kodaker Leonard Holcomb gave us a 1980 desktop computer and hundreds of discarded five-inch floppy disks. Even by the standards of 1991, this computer was a dinosaur. There was no hard drive, and it required a floppy disk to boot up. It had sixty-four *kilobytes* of RAM. When you flipped on the power, the fans whirring inside the twenty-pound beast made you believe it was going to blast into space. Even though my family had no money, I had a computer—such as it was—before most other American households.

The machine was a gateway to a new universe. I would pop in a random disk and enter the name of every executable file on it. Some of these programs were rudimentary accounting software—*boring!*—but many were games. Most of them were the creations of hobbyists. Some were text adventures; others had primitive animations that played on the monochrome monitor already burned in by years of use. The display was green pixels on a black screen, with no shading possible; the pixels were either on or off. The machine was limited in other ways too—namely, no mouse, word processor, or modem.

Sensing my burgeoning interest in computers, my father's sister, Aunt Marie, in 1993, gave my parents money to buy an upgraded machine. Aunt Marie, my godmother, worked as a housekeeper and could barely afford the expenditure herself, but she knew what the gift meant to me, in particular.

It was an act for which I could never adequately express my gratitude. There was so much information locked inside these computers, so much to learn, so much to find out. The games were endless, and you could spend a whole lifetime learning. My Aunt Marie provided me with an outlet when I needed one badly.

My early memories of home movies, computers, and even taking a stab at programming were happy and productive, the kind of nostalgia-infused memories that remain sepia-toned. If the situation with Andrea in my childhood home was unsteady, these creative outlets were grounding.

In the late 1980s, when I first became fascinated with audiovisual technology, *Star Trek* was barely twenty years old but already felt timeless. I watched the original series every afternoon on our local Fox affiliate. Eventually the station stopped airing it, which for me was a small crisis. In 1994, I wrote a letter—on paper, sent in the mail—to the station's director of programming, asking if they might run through the episodes again so I could record them. To my amazement, she wrote back: I was in luck. That summer they'd broadcast every episode, one per night, beginning at 2:30 a.m. I learned how to program the VCR to record all the episodes. This was a time that preceded Netflix and even DVR. If you wanted to watch a show on TV that aired while you were sleeping, you really had to work at it. Early in the morning I would return home from my paper route and watch the episode, enthralled, before anyone else in the family was up.

Science fiction and science became conflated in my mind: lightsabers, light speed. I gobbled up books about biology and physics and watched documentaries like *NOVA* and *Nature* on PBS. I had two sets of encyclopedias, both hand-me-downs. One was a set of the *World Book* from when Jimmy Carter was president; the other, much older set was from a forgotten publisher, produced during the Eisenhower administration, which I kept at my bedside. In these explorations I unearthed the first kernel of who I was: the scientist.

In the late 1980s and early 1990s, it was easy to dive headfirst into science. It was an era of renewed interest in space. NASA was everywhere. How far could humans go now? The Strasenburgh Planetarium in Rochester was hallowed ground. It had a 1960s-era Carl Zeiss star projector that looked like a two-ton praying mantis. To this day, once per month, the planetarium eschews modern digital programming and unearths "Carl" from a well in the floor to immerse audiences in the midcentury sky. What lay before us felt infinite, and for the explorer class that meant a wide world was open to us. So much out there, so much to examine, so much to pull apart and put back together.

My scientific interests extended from the cosmological to the molecular scale. The latter interest was stoked by two years in the classroom of Mr. Tim Kearn, who taught me chemistry. The topic really clicked. I was good at it because I was able to imbue the elements with humanlike characteristics. In fact, the pioneers of the discipline seemed to have had a similar bias toward anthropomorphism: "excited" states, "noble" gases, and "frustrated" electrons. In contrast to chemistry, physics was too abstract for me, and biology involved too many terms and structures and too much memorization.

At the same time as I was excavating the science-minded side of myself, I was exposed to music, which has been a constant part of my life since the age of six. My mother figured I was interested in music because I was always humming the music from Nintendo games, so she signed me up for piano lessons. There, each Saturday morning for two years, the stern Mrs. Parsons taught me the half note, the quarter note, and how to read the treble and bass clefs. We got through the first two volumes of the Leila Fletcher beginners' piano series. I don't recall being made to practice by my parents, so I had some degree of self-motivation. Nevertheless, I didn't develop a strong affinity to the instrument until much later, well after I had stopped taking lessons.

In the fourth grade we were allowed to choose an instrument to learn and join the school band. I chose the trombone, and my relationship with music was reignited. Once again I fell in love with music, and this time my love affair would last. For a while I was merely a decent player, but when high school started I was faced with others who were better than I was. I was placed third chair in the less-serious of the two bands—the Concert Band—at Hilton High School. Granted, this placement owed to my being a freshman, when there were no freshmen at all in the trombone section of the coveted Symphonic Band.

I spent a lot of time practicing. I was fueled by a fire, even if that fire was merely the desire to outperform my classmates. I wanted to be the best. No, I *needed* to be the best. I skipped lunch. I skipped gym. Hilton had an exceptional music program and sent many students to the Eastman School of Music, the University of Rochester's world-class music conservatory. My practice paid off. I made it to Symphonic Band and was also selected by its director, Gina Ferraro, as first chair.

Science was still there in the background, but I hadn't done any extracurriculars in science since Mr. Lamont's REACH program in elementary school. Band was consuming. Music was more social than science, and I practically lived in the band room. I even picked up the piano again, taking lessons with my former middle school band teacher, Jon Yates. My progress as a musician felt measurable: I could play faster and cleaner just by putting more time in. By the time I was halfway through high school, I had a new trajectory in mind for myself: I wanted to be a musician. I wanted to be the principal trombone of the New York Philharmonic (a position, for what it's worth, that has been held by Joseph Alessi since 1990, when I was just seven).

It seems a little foolish now, thinking back on it, that a teen version of me thought he had everything figured out. I was going to be the next Joseph Alessi! But it was, too, a sign of my own dedication to the craft. Still, my passions and persuasions were cemented young. Computers and chemistry bore into me a love of one thing, *Star Trek* a love of another, the trombone a love of a third thing. This trifecta helped carve out a niche for who I was, and I had started to imagine a life for myself outside of Hilton—and that life was exciting and full of possibility. I had a lot of big, impossible dreams, like any kid, I suppose, and I was excited to set those dreams in motion.

It would be exaggerating to say that science had been placed on the back burner and truer to say that I had tacked toward music in terms of life ambition. To become a top-tier professional musician, I had to set my sights incredibly high throughout high school. When I was sixteen— probably too late—I managed to secure, through a family connection, private lessons with Dave Richey. As a thirty-year veteran of the Rochester Philharmonic Orchestra, Richey was a regional celebrity in the trombone world, and I drove to my lessons with him every week for two years. Kind but firm, Richey insisted that I know the slide positions and intonation for every etude assigned to me, asking me to start over after every error. Then, he could do his work of teaching me musicality and advanced technique. Easier said than done. Nevertheless, I got quite good, making first chair in several regional honors bands.

Determined in my course of action, I applied to two colleges for music: Eastman and Northwestern, both renowned for, among other things, their trombone studios. My high school jazz band director, Rick

Trott, admonished me for applying to only two music schools. I told him that it was best or bust and that if I wasn't good enough to get into a top conservatory at seventeen, I'd never be good enough to get the job in a top orchestra. He was dismayed but didn't press the issue.

My mother and father, the opposite of modern helicopter parents, never weighed in on my choice of where—or for what—to apply. They trusted my judgment and had no idea what it took to get into a top music school. The University of Rochester—of which Eastman was a part— along with Northwestern University were both excellent STEM schools, and I applied to both as a double major in trombone and biomedical engineering.

My auditions were disastrous. My parents accompanied me to Northwestern for moral support and because they wanted to see Chicago. It was the only college visit of any type that either they or I went on. I was to play my audition in the oaken studio of Michael "Mick" Mulcahy, the renowned trombonist of the Chicago Symphony. What I heard through Mulcahy's door was both awe-inspiring and completely gutting. Player after player went in the door, blasted out *Ride of the Valkyries* and a few other excerpts with extreme virtuosity, and emerged triumphant. Just as Ms. Ferraro had warned me, these trombonists ate, breathed, and slept music, just as I should have been doing.

When I went in I gave Mulcahy a saturated handshake.

"There's no reason to be nervous," he said in a kind Australian accent.

Asking my teenage self not to be nervous was like asking a candle not to melt in the flame. The trombone's slide amplified my trembling, making it impossible for Mulcahy not to notice. When I was done trying to pass off my nervous honking sounds as music, Mulcahy thanked me for my time and complimented me on my beautiful horn.

My audition in my hometown, at Eastman School of Music, was equally bad. The august John "Doc" Marcellus, former principal trombonist in the National Symphony Orchestra, also commented on my nerves.

When neither program accepted me, I refused to lower my standards. I was stupid, snobbish, and elitist as a seventeen-year-old. I thought to myself, *I'm not going to go to SUNY Potsdam or Fredonia to become a music teacher.* Ironically, becoming a teacher is exactly what I ended up doing.

I had to reconsider my college plans. If music wasn't where my life was headed, what was next for me? The next obvious direction, of course, was science, my other enduring love. The immediate question was, Where should I go to college?

It didn't take me too long to come up with a decision. The only vacation I had ever gone on with my family had been to Boston, to visit my Aunt Marian and Uncle Mike. So when I was accepted to Boston University, it felt at once familiar and like a good fit. I never visited the school, but they offered the biggest financial aid package of anywhere that accepted me. In 2001 the tuition and room and board were $42,000 per year. I received $24,000 in the form of a scholarship awarded for both need and merit, and $8,000 in federally subsidized loans. My parents paid the last $10,000. This amount represented nearly all their discretionary income. Deena had already graduated from SUNY Brockport, and my father was receiving his Social Security benefits.

I had a romanticism about Boston, the birthplace of the Revolution, the city on the Charles, with its leagues of academics and historic buildings bedecked in ivy. In some ways my feelings about Boston never actually changed. Something about the city and its heavy concentration of academics only solidified my belief that it was the right place for people who took themselves seriously.

Heading off to Boston, I saw myself as moving into the world of knowledge, pursuing the path forged by my mother as opposed to the path set by my father. In 1949 he had been accepted to Northeastern University in Boston for math but instead had taken over the family tailoring business to support my grandparents. My parents drove me to college in my 1991 two-door Buick Skylark, because my father's Mazda had 250,000 miles on it and may not have made it to Boston. But we were going, at any rate, busting out of Dodge, making moves toward that shining city on a hill. Finally, I was getting out.

About forty-five minutes into our trip, right after we got on I-90, I saw a young deer on the side of the road. I tried to alert my father of the deer, which had started to trot into our lane, but it was too late. We hit the poor animal, still spotted like Bambi, and it flew into the air, its body spinning like a boomerang. We had been going sixty miles per hour on the highway, and the deer landed smack on the median. My father, after pulling into the shoulder, fumbled around for a nitro patch to prevent himself from having a heart attack. A minute later a state trooper was there, pulled up behind us.

"What are you folks doing on the side of the road?" the trooper asked.

"Jesus Christ, did you see what just happened?" my father said. He had a way of saying things in moments of emotion, with a nasally voice and up one octave. Here was my father again, huffing and puffing and making a bit of a scene. It was far from the glorious send-off to a new life that I had anticipated when I had signed up to leave town for school.

The officer walked over and verified that, yes, there was a dead deer on the median. Thankfully, he didn't have to shoot it. He sent us on our way, just another day in the life of a state trooper.

The next morning we drove up to Kenmore Square, the eastern terminus of Boston University and steps from Fenway Park, home of the Red Sox. I found myself instantly overwhelmed. Cars were honking. The Green Line trolley that runs the length of Boston University was intimidating to a small-town boy like me. My freshman-year dorm was also located in Kenmore: a converted hotel called Myles Standish Hall. From my room on the seventh floor, you could see the outfield of Fenway. Boston's famous Citgo sign was two buildings over and eye level with our dorm window. It lit up our room at night in an oscillating pattern of red, white, and blue.

Myles Standish Hall, since renamed 610 Beacon Street, was shaped like a narrow wedge, crammed into the fork created by Beacon Street and Bay State Road. This narrowness created some interesting room layouts. My friend and political sparring partner Scott lived in a crevasse the width of a twin-sized bed that we called the Scott Slot. I was assigned to a double with my roommate, Dave, a mechanical engineering major. Dave would be my roommate during all four years of college.

Dave and I lived in a suite with a bathroom shared with two other guys, who each had a room to themselves on either side of ours. On the left side was Wilson, an international student from Indonesia. Wilson was a master of computer hardware and a lover of scotch. On the right side, literally and figuratively, was Greg, an intensely funny electrical engineer and George W. Bush–loving conservative. Back in Hilton most of the people I had grown up around had been conservative too, so Greg and I got along well. In my youth, conservative politics were all that I knew. Arriving at college, I also considered myself a conservative.

Hilton wasn't rural, Bible-thumping Republican—the kind of Republican that wades into the social matters of abortion and gay rights. It was a town driven by economic principles. A town left behind, in some ways, by changes to farm work and loss of industrial jobs, for example, at Kodak and Xerox. In shifting economic times, it made sense that some voters related to conservative values. The teachers at the public school, from what I knew, were Democrats. I was proud of my conservatism, but I also didn't entirely understand it or its historical roots. It felt a little subversive. I was entertained by the rollicking nineties conservative echo chamber, by people like Rush Limbaugh, who my father listened to on the radio in his tailor shop. Republicans in the media had marketed themselves as anti-elite, even though the Republican Party could count on the support of many individuals in high-income professions.

Politically, my mother was a conservative Catholic for most of her life. She was an artist, but she was never fully on board for the artist's lifestyle, the bohemian aspect of it: weed and free love and the bloom of the 1960s coming alive. But I respected that that lifestyle wasn't for her. Conservatism was what she had known, socially and politically, and she adhered to it until, eventually, she didn't.

My mother, for instance, admired Carter. There was something about him that stuck with her. Maybe it was his kindness and faith, which appealed to hers. She once took a photograph of him from across the top of his limo when he came to downtown Rochester. It was a good picture, albeit taken in haste, as a crowd assembled around the president. "Don't take my picture!" a Secret Service agent snarled at my mother in a bizarre overestimation of his importance. The large photo still sits in my mother's family room.

Over the years my mother started to perceive a vein of cruelty and anti-Christian behavior normalizing itself within the ranks of the Republican Party, and it was something she couldn't abide.

Even though my mother was conservative, she had a box of books in the garage, books that did not seem to comport with her politics. They were all yellowed paperbacks from the sixties: *In the Shadow of Man* by Jane Goodall, and a bunch of lefty novels. Books by people like Aldous Huxley, people who would go on to inspire me, to force me to think differently.

Oddly, my father was a Republican while the vast majority of Italian immigrants and Catholics were Democrats. He was the only Republican in his nuclear family, a fact that irked his sisters as he sparred with their Democratic husbands. When he moved to Hilton, he fit in, politically, at least. He was a small-business owner and adhered to the ethos of pulling oneself up by one's bootstraps. That conservative sentiment—the sentiment held by my parents and, largely, by my town—had been bred into me. My conservatism had lived in me, then undisputed, uninterrupted, until something happened at the very beginning of my first semester in college, in September 2001.

Chapter 3

The first thing that happened during the start of my freshman year was that my computer crashed. Maybe this doesn't seem like a story worth retelling, but the crashing of my computer set in motion a series of secondary events.

I hadn't been able to afford a top-rate Gateway or Dell. Instead, I bought a machine from a dealer at a computer expo in Syracuse. It was a bargain-basement hodgepodge of parts. That piece of garbage just couldn't compete with the workload I had forced upon it. Once, playing the grotesque shooter game *Soldier of Fortune* fried my motherboard. I took this giant computer to the campus repair shop, hopeful that the university geniuses could work their magic. There, amid the parts and the dust and the people working away, I noticed that the television was on. I couldn't see what they were watching behind the counter, but it felt important.

Oddly, no one mentioned the television in the background. No one mentioned anything but my sad, scrap-metal pile of a computer.

"We can't fix this," they said when they saw the computer. "It's a show computer. There's no warranty." The television hummed in the background. No one said a thing.

Defeated, I started to walk back toward the dorm. I began to catalog my growing list of problems, which started with the fact that I was now a freshman without a computer. That's when I heard someone say something alarming.

"New York City's on fire," a student said. "Something has happened."

It took a moment for me to register the words. How could the city be on fire? What could have happened during my short trip to the computer shop? What *had* they been watching in the repair shop anyway? I replayed the moment in the repair shop in my mind, tried to zoom in on the television that had provided background noise. Nothing. I couldn't see it in my mind's eye.

Instead, I went back to the dorm. There was no cable, so we got the feed on my mother's thirteen-inch TV, which I had hauled to Boston. I had made a poor man's rabbit-ear antenna from a coat hanger needled through a safety pin, which I had jammed into the coaxial input on the back of the TV.

By the time we turned the television on, the first tower of the World Trade Center had collapsed. A landmark—something that had defined the city's skyline for my entire life—turned to dust and rubble. Twenty minutes later the second one was gone. People were jumping from buildings. The same footage replayed on a loop: firemen rushing in, the towers collapsing into billowing clouds of black smoke, people running and screaming on the streets of New York. America was experiencing a tragedy so great that I knew we would never be the same. My roommates and I couldn't pull ourselves away from the television set.

Everything about my life was different after September 11, 2001. It was my generation's JFK moment. Most of my peers can remember exactly where they were—exactly what they were doing—when they first learned of the tragedy of the towers. There was the moment Before and the moment After, and those moments can never be reconciled. The safety of life before 9/11 will never be replicated. Air travel will never be the same again. A fear crept over us. Were we safe in our major cities? Airplanes? Could we trust our fellow Americans? The serenity of our country had been disrupted in a major way, and we had to live with the ugliness now, not just for a little while but forever.

I felt like I had become an entirely different person, and not in a gradual way. My metamorphosis had been instant. It had taken place in just seconds and felt profoundly ungrounding. When kids go to college, they have an adjustment period. They learn, change, and grow. I had to become an adult in just a few minutes, when two planes crashed into two buildings I felt like I had known my whole life, even though I'd never been to New York City. Nevertheless, life did look different after that: less about what I wanted to become, and more about why any of it mattered.

My politics and my belief system became unmoored. After my rejection from music school in the spring, I had steeled myself for studying engineering. The natural sciences felt too pedestrian and possibly too easy. Not chemistry but chemical *engineering*; not biology but biomedical *engineering*. I had chosen biomedical engineering not because I knew what it was but, rather, because I had believed that it sounded cool. The program included a smattering of courses in every major STEM subject: biology, physics, math, chemistry, and computer science. I know now what kids with an affinity for science and engineering don't know, which is what defines the different engineering majors, unless they have a close relative in the field. To me, biomedical engineering was as wild and mysterious as bionic limbs. It tapped into the same part of my brain that had ignited during those late-night *Star Trek* recording sessions. Just as I had been sure of my passion for music during my years of high school, I had convinced myself that I was passionate about biomedical engineering. After all, it was what I had come to Boston University to do.

After September 11, biomedical engineering no longer seemed as important. Faced with an existential question at eighteen—*Why do people behave the way they do?*—I dropped my engineering courses and changed my major to anthropology. I wanted to study human societies in the dispassionate way that aliens might. To excavate what motivated the kind of evil I had experienced during my first days of college. The kind of evil that could sweep through my country and extinguish lives. I just wanted to understand it all, to become immersed in the workings of human beings.

I didn't give up on the natural sciences. For one thing, I chose the biological—as opposed to the cultural—track of anthropology, the materialistic branch of the field concerned with fossils, primates, behavior, and evolution. I remained in my chemistry class because it felt important too. I didn't really have a word for what I was experiencing. Maybe that word, in retrospect, was *epiphany*. I enrolled in an anthropology class, which was about human evolution and behavioral biology. The class was taught by Laura McClatchy, who assigned *Lucy's Legacy: Sex and Intelligence in Human Evolution* by Alison Jolly. It was about primatology and how clues from the past inform how society is in the present. Jolly's book led me to Edward O. Wilson, to biodiversity and conservation, to theories of how science and other disciplines are connected from physics all the way through music. It's a historical accident, Wilson posited, that fields are divided into categories. What is chemistry but emergent phenomena from physics? What's biology but emergent phenomena that arise from chemistry?

During my second semester I met a brilliant theoretical chemist, Professor John Straub. Sitting in class one day, I thought, *This guy gets it.* There was something sensitive in the way that he spoke to the class, and to me. One evening I met with him, eager to unpack what it was about him that spoke so deeply to me. I'm not sure what compelled me to stop by his office, and it hadn't occurred to me that it might be strange for a professor to be in his office after dark. Nevertheless, there he was. We talked about the chemical origin of life, and about how I didn't have to drop out of the natural sciences to retain my humanity or my natural impulses or my curiosity about human behavior. We also talked about his research—pure theory, "pencil and paper," he called it—which was not my cup of tea. However, I was fascinated by the goal of his field: theoretical biophysics, or the use of physics to predict the behavior of molecules and cells. In principle, Straub sought to understand how simple things like the attraction of positive and negative charges gave rise to the complexity of life.

Straub had a suggested reading list for me, which included, once again, books like *Lucy's Legacy*. I read his suggested books, intrigued by the fact that chemistry was at the base of all of it. One of his suggestions, *The Selfish Gene* by Richard Dawkins, put forth the argument that all of life and its complexity—including, ultimately, society and culture—arose from competition between genes. Genes are, after all, just long strings of DNA, molecules that obey the laws of chemistry. This was it, the connection between the sciences and humanities for which I had been searching.

I talked to John Straub about changing my major again, this time to chemistry. Something had been peeled back through all of this, some greater understanding revealed. If I had arrived at college a little green and unaware of the possibilities open to me in the world of science, my limited time there had already opened doors and intellectual possibilities. All of a sudden the collision of world events and professorial guidance had offered me new challenges and ways of seeing the world.

Diving into these interests, I suddenly became aware that I no longer fit in with the type of conservatism that had defined my upbringing and cultural milieu. Back then Rush Limbaugh and Bill O'Reilly cornered the market, as far as right-wing thinking was concerned, and their talking points were simply not compatible with the bigger ideas that I was struggling to process. They indulged in folk science, which made me bristle, even as a conservative teenager. For example, during a Clinton-era controversy around logging rights on federal land, I heard Limbaugh suggest that the endangered spotted owl simply find a new habitat. ("It has wings, it can fly!" he bellowed.)

In the aftermath of the greatest tragedy to take place on American soil since Pearl Harbor, it felt important to look beyond personal grievance and toward something more potent. Here I was, trying to parse human sociobiology, the genetic reasons behind social behavior and game theory. On the other side, there were these antagonists with very basic ideas that suddenly felt unchallenging and flat.

My burgeoning interest in anthropology had opened doors for me, but I was still somewhat rootless. I knew I wanted to do undergraduate research. However, there was no prescribed way to do it—and there still isn't. Sure, you can knock on a professor's door, but they don't have to answer it. They don't have to respond to your emails. They may not have space for you. After all, it's not going to be the professor who will be mentoring an undergraduate in the lab. The majority of professors at R1 institutions—which encompass the top research institutions like Harvard, MIT, and Stanford—rarely set foot in the lab. Maybe they go in a few times a month. In the most elite places, they may go in only a few times a year.

The de facto job of a professor, for what it's worth, is to generate ideas and to secure funding for their research (e.g., by writing grant proposals to pay their students' salaries). They must also make others aware of their work. They read the literature in their field and collaborate with their students to design experiments and to write papers. They also review manuscripts and grant proposals written by other researchers (i.e., peer review). Professors travel extensively to present their results and learn the latest and greatest from others. Sometimes this travel involves outreach in international locations. For big-shot professors, these trips take place multiple times a year. In the hazy days before I had my daughter (the same hazy days before COVID), I logged 110,000 air miles a year, achieved Premier 1K status on United Airlines, and came to expect lie-flat upgrades on international flights. Oh, the disappointment when I ended up in economy instead!

As an undergraduate sophomore, however, I was starting out on the bottom rung. I needed a way into research, a hook, to bypass the random door-knocking and cold introductions. I had to command the attention of the professors who were busy with their trips and their grants. There was a program known as the Beckman Scholars Program, created by the inventor of the pH meter, Arnold O. Beckman. Every year, three students would be awarded a fifteen-month grant, which included two paid summers ($5,000 each summer) and two academic years ($3,600 each year). It even included travel to the headquarters of the Beckman Foundation in Irvine for an annual symposium—I had never been to California!

I was grateful for the opportunity to apply for this scholarship. The chance to get paid for doing research as an undergraduate is exceedingly rare. Lower-income students in STEM disciplines often face a choice:

work on campus shelving books or making lattes (which covers living expenses) or volunteer in a lab.

One of the professors affiliated with the Beckman scholarship was James Panek. Panek is an organic chemist. The word *organic* is commonly misunderstood. While originally intended to refer to molecules derived from biological sources, it has come to mean any molecule whose skeleton is made of carbon atoms. Ultimately, even these are derived from biological sources, even if the organisms that produced them have been dead for millennia. These dead organisms are converted over millennia into carbon-based fossil fuels and chemical feedstocks. Panek was an expert in a subfield of organic chemistry known as synthesis. When a drug molecule is discovered in some bacterium or sea sponge, it's ground up and separated into compounds and tested. The hope is that some of these compounds kill pathogens or cancer cells or have some other medicinal benefit. If one of these natural products is a hit, it is the job of a synthetic chemist, like Panek and his students, to figure out how to make these compounds from simpler starting materials.

I had listed Panek as my first choice of mentors on my application to the Beckman program, but I started to have second thoughts. He was a hard-driving guy who demanded that graduate students and postdocs work a minimum of twelve hours on weekdays and five hours on Saturdays. He didn't insist on Sundays, though his students came in anyway.

"I have a rule of six experiments per day," Panek had told me the first time I met him. To do this, you had to work well into the night or wake up super early. I was in the lab for over two years, and at most I only ever did three reactions in one day. One or two were more typical, and that cost me five hours per day as an undergraduate researcher who had a full slate of classes.

After an intimidating interview with a panel of faculty members, I learned I was a recipient of the scholarship, which was great news. However, Panek's intensity scared me. I asked if I could choose another professor with whom to do my research, but the committee was firm. I had to choose someone who was affiliated with the Beckman program, and Panek was the only organic chemist.

Once we started working together, Panek and I hit it off easily. He was from Buffalo, an hour west of Rochester. He described to me once having been weaned on Genesee Cream Ale, a recognizable brand from Rochester. He liked grunge and progressive metal and had a soft spot for my favorite band, Tool. He told me once that he owned the album with the cover that had a photo suggestive of a contortionist "going down on himself" (that would be the band's 1996 release, *Ænima*). Graduate students had seen him at concerts. He was an everyman with a gruff exterior, and everyone called him Jim, even his undergraduate researchers. He was also a sharp scientist who could turn it on when he wanted to.

Every two weeks we had a subgroup meeting during which we went up to a chalkboard and drew the molecules we were trying to make and the reactions we were trying to execute. One time I presented my research in one of the shopworn flannel shirts I had obtained as a teenager, out of necessity, from the church basement. He offered to buy me a new shirt.

As a mentor Panek had great practical suggestions. He also had a streak of perfectionism. The chemical structures he drew on the board were impeccable: Every chemist has their own handwriting depicting zig-zagging bonds, ring-shaped molecules, and electron movement. Panek's drawings were perfection, and he had similar expectations from his students, even the undergraduate researchers. My project, he wanted me to know, would require a lot of my time—about twenty-five hours a week. In return, he told me, I could expect to get into the graduate school of my choice.

I was simultaneously elated and overwhelmed.

My day-to-day work—creating a molecule—would eventually lead to a peer-reviewed scientific publication, one in which my name was listed first. The "first-author paper" is currency in scientific culture. It's a notch on the belt. No one can ever take it away from you. It follows you wherever you go. Each paper has multiple authors, and the first-listed author is the person who did the lion's share of the work. The final-listed author is the professor: the person who runs the lab and who most likely wrote the grant proposal that funded the work. The professor receives an asterisk, which signifies the corresponding author or the point of contact, because the first-listed authors are students.

A professor has many jobs. Jim Panek, for instance, taught one class per semester, and he also ran his research lab. His duty as a lab director was to mentor students and—critically—to bring in as much grant funding as he could and publish his group's research in journals with high visibility and impact. Panek was the facilitator, making the production of papers possible. Students like me did the legwork, grinding out the results. In research, the role of a professor is the principal investigator, or PI. In a grant application, PI means professor. Similarly, in the hallways of university buildings, a student adopts this lingo, saying "my PI" rather than "my professor."

Every lab has a superpower, a thing or collection of things it does better than others. For example, Spider-Man has super strength, sticks to walls, shoots webs, and has his Spidey sense. Similarly, a chemistry lab might specialize in a particular catalyst, make a particular type of molecule, or have an overarching hypothesis that guides its selection of projects and design of experiments. These superpowers are built on by generations of students, one paper or PhD thesis at a time. Discoveries can arise from the validation of a theory or by an incidental observation of an unintended or unknown effect—in other words, by accident. The intellectual history of each academic lab has the shape of a tree, as some lines of inquiry can birth others. The whole structure is documented in the form of papers, the corpus of work that becomes the group's legacy.

My first project as a member of Panek's lab was on a molecule called basiliskamide. In my first ever presentation to the other researchers in my group, I proclaimed—not untrue—that "the molecule was named after the basilisk, a mythical chimera that had the body of a serpent and the head of a cock." Laughter erupted. "*Rooster*, Darren, *rooster!*" admonished one of the graduate students. Basilisk*amide* was made by a bacterium living within a sea sponge off the coast of Papua New Guinea. It was named after the basilisk by the marine chemists who discovered it, in reference to a rock formation on the beach near where the sponges that harbored it lived. The molecule was purported to have therapeutic activity against *Candida* and *Aspergillus*, types of fungus which are responsible for everything from diaper rash and yeast infections to deadly growths in HIV/AIDS patients. So basiliskamide was a type of antibiotic compound called an antifungal.

Ultimately I successfully made basiliskamide from scratch. Comparing the spectrum of the compound I made to that of the material that came from the Guinean sea sponge proved that what I made was true basiliskamide—a complex natural material made synthetically, by me, from small building blocks. I published my paper in the fall of 2004, in the journal *Organic Letters*. It's unusual for an undergraduate to publish a first-author paper. Executing the experiments and iterating through failure is only part of the challenge. A scientific paper must be meticulously documented. Every molecule you make must be examined by at least five different techniques to prove its identity. It was a challenge.

Things were stacked in my favor. I had a full hood to myself that, like a ventilation hood above the stove in a commercial kitchen, allows one to work with chemicals safely. I had all my own stir plates, vacuum pumps, and solvent bottles and a sink and desk in the lab, which was unheard of for undergraduates. Most importantly, I had an attentive, invested mentor in Niall Langham, a senior PhD student (now a senior scientist at a pharmaceutical giant). Niall designed the synthetic plan— like a puzzle—that would eventually get me to basiliskamide. Despite my apparent success in the lab, Niall often expressed frustration with me. He found me arrogant. I'm sure I came off that way, though I was just insecure. I hid my insecurity by projecting confidence. I talked about my grades, about my performance, about how I thought I had aced a test. Niall was right to be annoyed.

Honestly, I knew what I was good at. I was good, for instance, at taking tests. I was good at the bookish part of learning. I learned, after publishing my paper, that I also had the obsessive qualities that make for a good researcher. I decided after just one year in the lab that I wanted to do a PhD and that I wanted to do it at Harvard. I felt like I had a shot. I had, at the time, one first-author publication and another in the pipeline. I had worked in the most famous chemistry lab at Boston University, and Harvard had someone who captured my science mind even more: George M. Whitesides.

Whitesides may be the most famous chemist in the world who has never won a Nobel Prize. He's more famous than most chemists who have. He's also the father of George T. Whitesides, a current congressman from Southern California and former CEO of Virgin Galactic, Richard Branson's company.

I first saw George M. Whitesides, the chemist, speak during the Beckman Scholars Symposium in Irvine, California, in the summer before my senior year, in 2004. I was inspired by his breadth of knowledge, his powerful baritone voice, and the way he carried himself. It was almost as if he were some kind of priest in the field of chemistry. He seemed infinitely knowledgeable and wise. I wanted to work for him.

Whitesides, to be clear, was more than just your average intellectual. He had founded twelve companies by the time I first heard him speak, and those companies had a market cap of around $20 billion. He was an outrageously successful faculty entrepreneur and led a large group of around forty grad students and postdocs. His lab had perhaps a $5 million annual research expenditure, mostly from grants—virtually unheard of in the field.

So I did what anyone hungry for a position might do: I introduced myself.

Chapter 4

Like most Americans, I had an *idea* of what it meant to earn a PhD, and that *idea* differed from the reality of what goes on in the actual world of academia. Those who attend college and end their education after graduation—or even those who pursue a master's degree and stop there—may not know what the everyday life of a PhD candidate looks like. To be fair, I didn't either. I believed, as many do, that the trajectory looks something like this:

1. Apply for various PhD programs.
2. Gain acceptance to PhD programs based on merit.
3. Choose a PhD program that is a good fit.
4. Spend several years in a designated lab, doing targeted and directed research that leads to an advisor-led thesis.
5. Defend said thesis before a board of advisors, experts, professors, and peers.
6. Graduate with a PhD.

In some ways, this timeline is accurate, but it also oversimplifies the process of the science-based PhD. Not all labs, for instance, are the same. Not all advisors are the same. Some students, I would learn, spend years researching a thesis project, only to learn that the project is a bust. The nature of scientific research is that some hypotheses turn out to be wrong. Inventions that work on paper don't work in the real world. Systems of complex parts often don't behave the ways we want them to. In the teeth of this truth, students and professors alike waste years trying to rescue their sunk costs. The exasperation is demoralizing to the point of breakage.

Students drop out.

Students "master out" (more on that later).

Students lose motivation, or experience friction with their advisors, and graduate in seven, eight, ten years, instead of the average of five or six.

As my time at BU ended, I knew only the rosiest version of what lay ahead. I held optimism in my heart. My hard work as an undergraduate was due to pay off. But would achieving my dreams get me all the things I wanted in life? Maybe there was no quantifiable answer to that question. I hoped the path I was forging would lead me to the place I wanted to go.

I didn't apply only to Harvard, because that would have been foolhardy. I still thought of the school as a goal that was a little out of my league. I applied to eight PhD programs: Harvard, MIT, Caltech, Stanford, Berkeley, Scripps, Columbia, and UCLA. I was concerned, in truth, that I would be accepted to none, as I had scored in the forty-eighth percentile on the math portion of the Graduate Record Exam (GRE), a true humiliation; I had a minor in physics, a math-centric discipline. I hoped my high scores on the analytical writing, verbal, and chemistry subject tests would outweigh my dismal math score, and I threw darts at the wall. My GPA had been strong at Boston University—3.71—which, before the days of intense grade inflation, was good enough for summa cum laude, the top 5 percent of my graduating class. I got a "with distinction" appended to my Latin honors because of my laboratory research, which produced two first-author papers. In the end, I was accepted everywhere I applied, with the exception of MIT. The outcome was beyond good.

For some this would have been a quandary. With so many excellent choices, how could I possibly have decided where to go? For many days I debated the merits of each place with my classmate (and academic competitor), Timothy R. Cook, who got into *both* Harvard and MIT. He chose MIT and is now a successful professor at the University at Buffalo and still a close friend. As far as I saw it, however, there was no real choice. From the time I set foot in Boston, I already knew the answer.

Harvard was its own city on a hill, where the pre-Revolutionary brick buildings and ivy embraced the deepest scholarship and knowledge. I knew even then that I wanted to be a part of that place. Stanford, Columbia, Caltech. They were all good. They were all *great*. But none of those schools felt the way Harvard felt, steeped in whatever magic and glory and old-school intellectualism I experienced when I saw that shimmering school while running around the Charles River.

Harvard had won my heart a long time before—this place of great thinkers, of great doers, of great feats of intellectualism. After all, E. O. Wilson taught there. I wanted to be a part of that place more than I wanted almost anything. It was a place where I could see myself achieving, where I could see the trajectory of my life changing. People went to Harvard to make their lives bigger, and I was no exception. The kid from Hilton was finally making it big. It was a dream come true.

When I started as a PhD student in the Department of Chemistry and Chemical Biology at Harvard, we were made to do three one-month rotations within different laboratories. I had gotten into Harvard because of my undergraduate research in organic synthesis, but I wanted to use the rotations to learn about other fields. Harvard had founded a new research initiative on the origin of life on Earth and other planets. Whitesides was part of this initiative. The basic research question was grand: how the conditions of the early Earth—four billion years ago— led to molecules that could replicate themselves, and how these molecules formed the first cells and proto-organisms. The first step in evolutionary biology, the field that so intrigued me because of the fire lit by *The Selfish Gene*—that is, molecules doing battle for resources and, as a byproduct, turning into life.

I thought to myself, *I should do my PhD on the origin of life*. The appeal went beyond philosophy. Discovering the conditions that might give rise to systems of molecules that could replicate themselves and adapt to their environments could have a real impact—for example, in the design of processes for environmental remediation and medicines that adapted to pathogens and cancer.

My first two rotations were in labs whose projects on the origin of life interested me on paper, but which had other problems. For the first rotation, I chose the lab of a PI who was talkative but extremely awkward, even by the standards of a Harvard professor. His students seemed to be as meek as mice, and there were rumors that the PI bullied them. The environment was uninviting, and I ended my rotation after one day. In the second rotation, the PI was kind and had a reputation for supporting her students. However, in my rotation I spent all my time testing a vacuum chamber for leaks. We never got to any actual science, and none was on the horizon.

I had saved my last rotation for the Whitesides lab. I was relieved to find that it had great camaraderie. A breath of fresh air compared to my first two rotations. It was nerdy, natural, and irreverent, though I noticed a curious combination of both admiration and cynicism in how they talked about their advisor.

Notwithstanding, this was an advisor who I had clamored to study under. I had specifically targeted Whitesides before I had even arrived at Harvard. He was a legend. The author of over 1,200 scientific articles and the listed inventor of at least 134 patents. According to the Hirsch index, he was, in 2011, the most influential living chemist. Back then, my ears pricked up at the mention of his name. And yet more than one student had something negative to report on him. So the question in my mind became: Why?

There was a joke that had circulated widely among the scientific community, always told in the same way.

What's the difference between George Whitesides and God?

God knows he's not George Whitesides.

During my rotation, I was assigned a fourth-year graduate student, Paul J. Bracher, to supervise my day-to-day activities in the Whitesides lab. Bracher is now a decorated faculty member in the Department of Chemistry at Saint Louis University. Even though Bracher started in the lab three years before me, we graduated the same year.

Like me, Bracher was also drawn to the Whitesides lab because of his interest in the chemical origin of life, abbreviated with the hip acronym COoL. However, despite Whitesides's purported interest in this research, he was slow to return Bracher's work. Months—and in some cases years—would go by without receiving any feedback. It seemed that Whitesides was more interested in talking about the origin of life—and in publishing pop-science articles on the topic—than in submitting hard manuscripts for peer review. As a result, Bracher had to all but abandon COoL research. As a matter of survival, he pivoted to a frankly easier project that commanded Whitesides's more immediate attention. Because of this pivot, he didn't publish his first lead-author paper from his PhD work until his sixth year in the program, a tragic deceleration for a brilliant young scientist.

"Joining this lab was the worst decision I ever made in my life," Bracher told me on the first day of my lab rotation. As far as I was concerned, the alternatives to joining the Whitesides lab were boring research, a less charismatic advisor, or a sterile social environment in which no one would dare say something as irreverent as what Paul Bracher had just said to me. I believed, perhaps simplistically, that if Bracher was having such a bad experience, he could have left and started over somewhere else.

What compensated for Whitesides's absenteeism as an advisor was that his research was interesting and dynamic. Whitesides had a ton of grant funding. We could buy any item we wanted, with no approval needed for anything under $500. If you needed to buy something for $5,000, no problem. You just had to ask the group's financial administrator, and approval was automatic. For the right kind of person, the lab was a playground.

The irreverence of Bracher and others appealed to me. I felt free to express myself fully and freely in the lab. No one was safe, even—or perhaps especially—our illustrious advisor. In fact, I used to do a fine impression of Whitesides. My voice, having dropped when I was eleven, dropped a few more semitones by the time I was twenty-two, and I was able to produce the timbre of Whitesides's commanding baritone. This impression nearly got me into trouble.

"Emily, I need your help with something," Whitesides once asked Emily Weiss as he walked into her lab and toward her desk. Weiss was an outstandingly productive postdoctoral fellow (and mentor to me), whom he often asked to prepare slides for conference presentations.

"What do *you* want, Darren?" Weiss sneered. And then she looked up, seeing Whitesides, who was perplexed. "Oh, it's you, George."

Whitesides shook off—and thankfully for my sake didn't inquire further about—her confusion.

Aside from the camaraderie, I really enjoyed my work. For five years I went into the lab seven days a week, unless I was out of town or writing a paper from my basement apartment in Harvard Square. I chose a research area that was generating papers more quickly than the origin of life. This area, which we called unconventional nanofabrication, involved developing manufacturing tools for making small structures. The thinking was—and still is—that future technologies will be enabled by materials with nanoscale dimensions—for example, all computer chips, including those used for artificial intelligence, along with devices for sequencing the human genome.

Most computer chips are made using huge fabrication facilities (fabs) that cost billions of dollars, take up acres, and are good for only a single generation of chips. The basis of chip manufacturing is a process called photolithography, from the Greek meaning the use of light to write in stone. In it, beams of extreme ultraviolet light are used to make transistors on wafers of high-purity silicon the size of vinyl records. These disks are then chopped up into thousands of chips—each containing up to one hundred billion transistors—that end up in our devices. It is the most sophisticated manufacturing process ever used for commercial production and has a very low tolerance for error, as defective wafers must be trashed.

The motivation for developing alternative (unconventional) methods of nanofabrication arises from the fact that some applications—for example, biosensors, solar cells, optical coatings for aerospace and telecommunications, and sensing of contaminants in the body or the environment—do not require perfection. These devices have nanoscale parts that can be made with cheaper tools and for far less money than computer chips. However, at the time I started my PhD, such tools were lacking. The Whitesides lab was in the process of developing tools based on miniaturizing processes that have been used for centuries: stamping, molding, and cutting. Whitesides wanted to know "how low can you go?" with these techniques.

I found this work fascinating, and accessible. I enjoyed working with my hands, learned a lot of materials science and the basics of electronics, and was inspired by the areas in which my work could be applied. It would become the basis of my research and, ultimately, my thesis.

Around the time I joined the lab, Whitesides hired two postdoctoral scholars, Emily Weiss and Ryan Chiechi. Weiss and Chiechi were interested in new types of solar cells based on nanomaterials. These solar cells promised to be cheaper than conventional silicon cells but were difficult to make. There was thus an opportunity to invent a new process to make this type of solar cell. The idea was that in a solar cell you need to have the materials that move around the positive charges and the materials that move around the negative charges be close to each other. There's a particular kind of solar cell in which the charges must be extremely close—twenty nanometers or so, which is a thousand times thinner than a single strand of human hair. My idea was that I was going to stack up materials, similar to the way baklava might be stacked, in layers. A hundred layers. And then I was going to roll it, like a Ho Ho from a truck stop, and chop it into slices. There it is: my solar cell.

The trick is that each layer, once dried, would tend to redissolve in the layer you coated on top of it. It would be a bit like painting over a bright red wall in fresh white paint. The liquid part of the white paint dissolves the red paint, and you get pink. Of course, paint is engineered *not* to redissolve, but our materials were not. So, to stack layers of films that could move positive and negative charges around, these films needed to be soluble in different solvents. One of the solvents was chloroform. It alone is nasty, neurotoxic stuff. If you've ever read a Nancy Drew novel, you know that one effect of inhaling the fumes is unconsciousness. The other solvent, methane sulfonic acid, though not volatile, is worse in one key way: It has the consistency of olive oil, and if you get it on your skin, it's difficult to wash off. And while it's on your skin, it's rapidly disintegrating your epidermis, nerve endings, and eventually other subdermal structures.

I had to filter my acid solution, shoving it through a fifty-milliliter syringe with a filter on the end that was fitted with glass wool fiber. It was just my luck: The filter popped off and sprayed acid everywhere, including all over my face and forearms.

I went first to the eyewash station to get the acid off my face and then to the bathroom to look for other spots. I took off my shirt, and I did my best to wash off the rest. I practically bathed in the elongated basin of the wheelchair-accessible sink, soaking my pants and everything I was still wearing. When I emerged from the bathroom, I was shocked to find that I was no longer alone. In the hallway a crowd had assembled. The building facilities manager had been called, along with the safety team. I was dripping, wearing no shirt and wet jeans, with acid burns still hot on my skin. A spectacle. Exciting in so many ways.

"Do you want to go to student health?" the department safety manager said. "Or do you want to go to the emergency room?" What had been a routine incident had somehow metamorphosed into a Major Medical Problem in a matter of minutes.

"I want to go to the emergency room," I said. I had no confidence in the student health team's ability to treat chemical burns.

"Well, for liability reasons, an ambulance has to take you there," the manager told me.

Despite the delay of having to wait for an ambulance, I acquiesced. Another student offered up gym clothes that had been stashed under a desk. I changed. The ambulance arrived. My wounds, it turned out, were superficial because I had washed them quickly. Nevertheless, I walked around for a month with what appeared to be chicken pox on my face and a nickel-sized scab on my forehead.

That was a lucky escape, a set of incidents that could have ended with a far different outcome. My pride took a hit, as did my vanity. I had just started dating the woman who would later become my wife, and my main concern, as EMTs assessed my level of pain, had to do with whether or not I would be permanently disfigured.

They seemed more intent on preserving my actual life, and the truth is, I had been guarded against damage by two things: first, the fact that I was wearing my safety glasses, which probably saved my vision, and second, my quick action in the eyewash station. Had I been wearing a lab coat, I could have prevented the burns on my forearms, the scars of which I still have. When examining the wounds on my arms, the ER doctor told me that she wasn't worried about my arms because they were "very hairy" and the hair would obscure the damage. She was more worried about my face, because I was an "attractive guy." Sixteen years later, I would commemorate this whole incident with a large tattoo of the solar cell molecule.

A long-lasting benefit of that afternoon's trip to Cambridge Hospital was the scare put into me by the triage nurse who measured my blood pressure upon admission: 170 over 90. "You need to watch your blood pressure," the nurse told me. Unfazed by my explanation that I was nervous because I was in the emergency room, she continued, "Your incident was over an hour ago. Your blood pressure should be closer to baseline now. There is evidence that hypertension in young people leads to early-onset kidney disease."

At the time, I was accustomed to drinking around three cups of coffee each morning. I discovered later by genotyping on 23andMe that I metabolize caffeine slowly. I didn't touch caffeine again for twelve years until the birth of my daughter, Cora, in 2019. I got off it for a while, then readdicted myself after I was prescribed Lexapro, as the SSRI was making me *too* mellow. I'm now in a good place, and so far nothing has ruptured inside my body. At least not anything that I don't already know about.

After the accident I was embarrassed to walk around the lab. I was also gun-shy every time I had to load even the most benign fluids through pressurized equipment. For the rest of my PhD, I wore not safety *glasses* but face-hugging safety *goggles* of the type worn by skiers and students in high school chemistry labs. Incredibly, in retrospect, I returned to work mere hours after my accident, in part to clean up the acid left on the floor and the walls but also to continue *working*. Whitesides found me later. Inspecting my wounds, he told me that my accident should serve as a lesson not to "force things." He was glad for my safety, no doubt, though he was probably also grateful that he would not be subject to personal legal scrutiny for the incident, since my injuries were so minor.

The Whitesides lab had last been renovated sometime in the early 1980s. It was outfitted with outdated cabinetry, all porous wood that had absorbed thirty-five years' worth of spills and smells. The lab, which took up most of the second floor in the storied Mallinckrodt Hall, built in 1928, was composed of ten approximately five-hundred-square-foot rooms, which each had its own signature scent. One day we had a lab clean-out, and I was assigned, alongside my friend Danielle, to clean out the organic storage cabinet. We had to be in there all day to update the chemical inventory for thousands of chemicals. So rancid was the smell that we had the door propped open. Seeing this, a fourth-year graduate student walked past and shut the door on us, sealing us in. "Don't prop that door open," she said. "It smells out here."

So there we were, closed in, enveloped in the smell, the toxicity, in the chemicals that could have (or maybe should have) been banned between the 1960s and the 2000s. Who knows? Like Count Rugen's torture apparatus in *The Princess Bride*, each hour in that stench probably robbed a year from our life expectancy. And we didn't push back either, since we were mere first-year graduate students—G1s, we were called, Harvard's word for *plebes*. If our senior labmate was intent on murdering us with toxic fumes, so be it.

One aspect of safety culture Whitesides could be counted on to support was the wearing of safety glasses. When he was a young professor, though already in his seventh year at MIT in 1970, his friend and colleague K. Barry Sharpless, who has since won the Nobel Prize (twice!), had a piece of glassware blow up in his face. He now has a glass eye. Was George, as we called him, going to get on our case for not attending our periodic university-mandated safety training? No. But he would admonish us for failing to wear safety glasses, even if he came into the lab only periodically during a five-year PhD.

When I tell people how little face time I had with Whitesides, I am often met with shock. Each PhD student and postdoctoral trainee presented their formal results in a group meeting presentation once per year for five years, and I personally spoke to him about my career trajectory perhaps four times. Getting an appointment to meet with him was like establishing care with a medical specialist. It wasn't uncommon to have to wait three to four months to meet with *our PhD advisor*. He was out of town so often that when he was in town, we would sometimes have group meetings multiple times per week, sometimes a day-night doubleheader, sometimes on Saturdays.

Whitesides's itinerancy necessitated one of the more infamous methods he used to direct his research lab—his handwritten comments on our printed "outlines." In fact, in 2004 Whitesides published a paper in the journal *Advanced Materials* on what came to be known, internationally and across disciplines, as the Whitesides method on the use of outlines to direct scientific work in a research group.

The outlines were the documents that would become the manuscripts we submitted to peer-reviewed journals. The first outline was to be written after we obtained the first result of any scientific project. A result was any interesting, new, or unexpected observation or potentially useful invention. This initial result was not expected to be a rigorously controlled experiment with a full complement of statistical analyses. When it had been fully fleshed out after a dozen or so iterations, Whitesides would write "I don't need to see this again," which meant that the outline had reached the status of manuscript and that it was ready to be submitted. This note, signifying completion of a project, and all others were written using his favorite tool, the Pilot Liquid Roller Ball pen, and the ink was *always* blue.

These blue notes in the margins of thick printouts, which Whitesides took with him on his travels, constituted most of the scientific mentoring we received on route to our PhDs, at least from our advisor. The legibility of these notes varied wildly. One was led to believe that some fraction of the time they were written in the cheap seats on a turbulent domestic flight when Whitesides's elite status upgrade didn't come through. (On more than one occasion Whitesides took meetings with students in the taxi on the way to the airport.) One time, another student and I were trying to decipher a particularly illegible blue scribble and determined that it could say nothing other than "I know you want to fuck it." Given the geometric constraints of fucking a single molecule, we gave up on its true meaning and chalked it up to turbulence. The collaborative and time-consuming task of interpreting the most illegible of the blue scribbles, according to Paul Bracher, was akin to deciphering the proclamations of the Oracle of Delphi.

To his credit, Whitesides once told us that, so exasperated by seeing the same types of errors on student papers over four decades as an advisor, he would write comments that "no good human being should ever say to another." This preemptive apology blunted the pain I might have felt upon one time seeing that something I wrote was "illiterate," or that another colleague felt about a comment in blue ink that his work was "a pig's breakfast." Ultimately, we laughed off such rhetorical excesses. Rarely did this discouragement spill outside the boundaries of blue ink on printed outlines and into real life.

I'd be disingenuous if, as a professor myself, I stood in too harsh a judgment of my PhD advisor for this colorful reaction to poor student work. I've certainly had colorful *thoughts* about student work that may have been rushed or careless, though I have generally tried to spin such thoughts into guidance the student might find actually useful. Social norms—and, yes, filters—are good things in professional environments. I can only hope I stick to mine when I've been in this job for thirty more years.

Many of us claimed to have wanted a more personal advising relationship with Whitesides, but the fact is, he scared me. I was always nervous in his presence. If I had known about beta blockers in graduate school, I would have sought a prescription. I often convinced myself that Whitesides would ask me a question I couldn't answer and call me stupid. Once he came close, writing in an email that a concept I didn't know off the top of my head was "idiot simple." Such admonitions don't sit easy with someone already afflicted by a pervasive form of insecurity called imposter syndrome. For me the feeling was not always without merit. I always considered my intellect like an overclocked microprocessor—an intrinsically slow piece of hardware that is made to run faster than its specifications by removing its guardrails. Such neuroticism produced a litany of psychogenic illnesses, including arrhythmias and sweating, along with my glasses fogging up during intense conversations.

Beneath my obsession with grades was a kind of panic: the sense that if I slowed down, even briefly, I would fall behind. My strategies for coping were not admirable so much as desperate. In my graduate course on quantum mechanics, the homework assignments were damn near impossible. I found a "secret textbook" at the bookstore that contained solved problems similar to the ones on the homework. It was a gold mine, and I shared it with my classmates, despite the large personal expense of one hundred dollars. One night we photocopied a big chunk of the book in preparation for the take-home final exam. The exam was open book, but I'm guessing the professor didn't know about the existence of *that* book.

These methods of gaming the system seem quaint by today's standards. In 2005 websites like Chegg—which contain databases of solved problems from nearly every textbook and every university—were just coming into existence. ChatGPT blew all that up. But for the grace of the divine, I was born twenty years too early to have been tempted as a student by omniscient AI tools. Homework may still be useful for the development of skills, but as a method of assessment, it's now useless.

I've heard it said that Harvard is a great place to *have been* as opposed to a great place *to be*. It's a cute saying, but the reality is much more nuanced. Even in the intense and overworked Whitesides lab, there were plenty of opportunities to celebrate and commiserate. The sheer size of the lab—over forty graduate students and postdocs—meant that there was always something to celebrate: people passing their exams and getting jobs, and all major holidays. Whitesides paid out of pocket to cater every one of these events. As long as someone else ordered and arranged delivery, he would give them his credit card, so to speak. The lab member who took ownership of the greatest number of these events was, paradoxically, Paul Bracher.

Despite Bracher's enmity toward Whitesides, he acted in ways to garner his approval, or maybe he took these leadership roles because nobody else would or would do them as well as he could. Each year, for instance, Bracher organized the annual gift. Whitesides and his wife, Barbara, used to host a holiday party for the group at their home in Newton, Massachusetts. They went all out. It was catered, complete with roving staff serving hors d'oeuvres and offering small napkins for which you never had a free hand. They hired a bartender, and wine and beer flowed freely. They recruited neighborhood teenagers to watch the group members' young children in the basement. Each year Paul spearheaded the creation of a homemade gift for Whitesides. One year the gift was a Monopoly set called Georgeopoly, in which all the tokens and properties were representations of projects in the lab. Another year: a large photomontage—a portrait of Whitesides where the individual pixels, if you looked closely, were small photos of past and current lab members, and images from projects. And yet another year, Whitesides's immortalization by the establishment of his page on Wikipedia. While Bracher organized these projects, nearly everybody contributed. Everybody's relationship with Whitesides was multidimensional. They were instructed by him and benefited greatly from being in and associated with his group, though many students left the lab, left the program, or left science.

It's possible that the strategy was put in place on purpose. I've heard it called the Academic Hunger Games—to hire more students than you need to advance your research. The rationale is that some will burn out, move on, not cut the mustard, or not get on board with your vision. In fact, this was advice Whitesides was said to have given to at least one postdoc who had recently won a faculty job at another university. While Whitesides's lab was perhaps the most famous lab at Harvard—despite the fact that three of his colleagues in his own department had won the Nobel Prize—it was not particularly difficult to get in, provided you got into Harvard in the first place. Most research groups at Harvard and similar institutions have maybe one to three slots each fall, mostly based on space and grant funding, to take a PhD student. Inevitably, some students who enroll at these schools to work in a particularly competitive lab are shut out and have to join the lab of their second or third choice.

On the other hand, Whitesides was not particularly discerning when it came to recruiting graduate students. As a case in point, I can't imagine I said anything as a trembling, sweaty first-year graduate student that could have remotely impressed him during our sole one-on-one meeting before he agreed to take me as an advisee. I remember telling him I wanted, someday, to write a popular science book in the style of *The Selfish Gene* by Richard Dawkins. He told me he was supportive of unconventional careers for PhD holders, but that even Dawkins started out as a serious researcher and that I should too.

I was a member of the Whitesides Group during the apex of its glory. Our group was publishing a peer-reviewed article approximately once every ten days, and it was during my time in the lab that Whitesides garnered his one thousandth publication. The lead author of this publication was my labmate, one year ahead of me in the PhD program, Eleanor Hutcheson. The paper described a high-throughput "worm clamp" capable of immobilizing hundreds of nematode worms on a small, transparent silicone chip. With Hutcheson's invention, the physiology of copious numbers of worms could be monitored in response to various conditions. Once let loose, unharmed, they could resume their wormy lives.

Yes, Whitesides is an organic chemist by training, but his thousandth paper was, bizarrely, on worm clamps. And that wasn't even the weirdest (or weirdest-sounding) project going on in the lab. Using flames to communicate. Using silicone rubber with liquid-filled channels as chips that could perform rudimentary computation. Measuring the conductivity of single layers of molecules using dangling drops of mercury. Devising methods of making water roll uphill. Modeling protein folding by putting static charges on the beads of necklaces. Making solar cells inspired by jelly rolls. The breadth of his interests and portfolio of projects was enormous. No other scientist has published with such authority in a broader field of disciplines, and it was an honor to learn from him, though we rarely did so in person.

In a typical PhD research group, candidates hold a standing meeting with their advisors every week or two. We got five minutes at the end of the annual presentation we made to the group (during the group meeting), held in the seminar room after everybody had left. The extent of these conversations was rarely deep or scholarly, at least with me and others with whom I spoke. He often commented on the design of our slides or the cadence of our presentation.

"You talk too much," he told me during one consultation, which I remember vividly. "You need to use *verbal punctuation*. You speak too quickly, and everything runs together. And you should walk around more when you give talks. It will make you more engaging to the audience." I recall feeling a bit cheated by this advice from the great man, as none of this advice was singular to Whitesides's vast expertise in my chosen discipline. I was never quite sure if he was paying attention to my presentation. I can only imagine the places to which his mind may have been wandering.

At my going-away party in the group meeting room, which happened upon the successful defense of my PhD, I was surprised that Whitesides heaped praise on me. "We're going to hear a lot more about Darren's accomplishments in the future," he said. "I'm devastated that his time in the lab is coming to an end."

I traveled to Cambridge in 2023 for Whitesides's eighty-fourth birthday and sixtieth anniversary of becoming a professor. This was a huge event, with alumni from across the decades in attendance. Without irony, the event was held in the exhibit halls of the Harvard Museum of Natural History. The aging Whitesides was recovering from a stroke and looked frail, though his voice was still authoritative and his handshake robust. He didn't remember seeing me since I graduated from the lab, thirteen years earlier. In fact, I had visited him several times, and he had been the keynote speaker when my team won Best Paper of 2018 in the journal *Chemistry of Materials*. I was happy to forgive the oversight.

In many ways the experience in the Whitesides Group was a distillation of what you'd find in any high-profile lab at an elite university. The groups are large and the funding is abundant. The professors at these places are famous—many are on an implicit quest for a Nobel Prize—and thus often absent, giving talks. The other students around you are bright and hardworking. Most of the day-to-day mentoring, it may surprise outsiders to learn, is left to the senior personnel, not the professor. As students in elite labs most likely already had intensive experiences as undergraduate researchers, they may in fact not need the advisor's direct mentorship to be successful. Perhaps not an ideal arrangement, but it's the way the system has evolved.

Chapter 5

How is success measured in science? There is what *really* matters—the importance of one's discoveries and inventions and the flourishing of one's students—and there is what is easy to measure: grant dollars, awards, invitations to prestigious conferences, and papers. Papers themselves are judged by the number of times they are cited, and scientists are often judged by the number of highly cited papers they have authored.

To make this judgment, academics often look to something called the h-index. The h-index, named for its inventor, Jorge E. Hirsch, equals the number of times papers authored by a specific scientist have been cited X number of times or more. If one's h-index is, say, 61, as mine is at the time of this writing, that means that sixty-one papers authored have been cited sixty-one or more times by other scientists. A scientist's total citation count—twenty thousand for me—may be skewed by a couple of outlier papers that have received an outsized level of attention. Mine is considered a solid level of productivity, but far from godlike: George Whitesides has an h-index of 300 and a citation count well over four hundred thousand.

In the gamification of scientific research, citation counts aren't everything. Some journals are more prestigious than others. The pecking order of the journals a scientist tends to publish in correlates with the perceived quality of their work. Getting a paper accepted to a high-impact journal is akin to getting an award for the research. Like elite colleges, the quality of journals is judged in proportion to the percentage of manuscripts it rejects. Besides the rejection rate, journals and the scientists who submit papers to them use another metric—*impact factor*—to judge quality.

The impact factor is the average number of citations earned by a paper in the previous two years. In my field, the highly respected *Journal of the American Chemical Society* (*JACS*) has an impact factor of 15, considered a high score, if not astronomically so. The threshold for an elite-level journal might be over 40. Such heights are reached by, for example, the big three: *Science*, *Nature*, and *Cell*. In these journals, the average number of citations garnered by each paper in its first two years is at least forty. A paper published in one of these coveted journals early in one's career can be a ticket to scientific stardom: for a trainee, a higher probability of a position at a highly ranked institution, or for a professor, accelerated tenure, invitations to present at top universities, and increased hit rates in grant proposals.

A journal may manipulate both the rejection rate and the impact factor simply by reducing the number of papers published. Increased selectivity increases the rejection rate, along with the visibility of the trickle of papers actually published. The availability heuristic thus increases the rate at which each paper is cited. This strategy is exploited to the maximum extent by the big three, which lend their names to "baby" journals, which carry nearly the same level of prestige as the parent journal. For example, while I have only one paper in *Science*, I have one paper each in *Science Robotics*, *Nature Nanotechnology*, and *Cell Metabolism*. There are other tactics even prestigious journals use to manipulate their impact factors. One example is coercive citation: the editors "suggesting" that the authors cite papers that *happen* to be in the same journal. Another is the publication of mini-reviews, perspectives, and opinion pieces, which tend to be cited a lot and require no original—that is, laboratory-based—research.

Journals' impact factors carry weight for scientists even if the papers they've published in them have never been cited. For example, in dossiers submitted for tenure and promotion, it is common for the candidate to list the impact factors of each journal in which they have published. They do this for the "benefit" of the tenure committee, who may not be familiar with the prestige of this or that journal outside of their own field. However, the effect of this practice is to take credit for the citations earned by *other* authors simply because a person had a paper accepted to the journal.

Citation metrics also tend to over-reward scientists who contribute small parts to big projects. For example, if I am the twentieth author on a paper with twenty-five authors, chances are high that I didn't do as much work as the first author—yet a citation counts toward my h-index the same as theirs. I also don't suffer much reputational risk if the paper has flaws. There is an open question as to whether every author on a paper must testify to the truth of all its claims, even if the claims fall within the expertise of only a minority of its authors. Peer review catches only the sloppiest of fraud cases.

Citations, impact factors, and h-indices: a scientist's version of hits, home runs, and runs batted in. This information is readily accessible on Google Scholar, and universities also have their own paid products they use for indexing purposes. To some, these statistics don't matter much. And, until 2012, they didn't matter much to me either. But that year I started applying for faculty positions, so my own personal baseball statistics did in fact start to become part of an important legacy (more on that later). A player is, after all, only as good as how they measure up against their teammates and opponents.

I was under no illusion that I would be as famous as Whitesides or my postdoc advisor, Zhenan Bao, Professor of Chemical Engineering at Stanford, who has an h-index over 200, with eight hundred publications. When she speaks at a scientific conference, it's standing room only. When her talk ends, her fans turn around and leave, to the chagrin of the student scheduled to speak after her. When I was a student, I remember thinking that research was a crapshoot: If your project worked, you would get a ton of papers and be set. If it didn't, you wouldn't. I had no idea where I would end up, but I had some goals. I wanted to have a productive research program and a few meaningful discoveries or inventions to my name, write books, and teach courses on topics that interested me. If I were really successful, I might get invitations to speak at companies or on TV. But before all that, I had to finish my PhD.

Some PhD candidates complete the research that will become their thesis in five years, some in up to ten. Some enter the program, discover that they can't—or don't want to—commit for the long haul, and finish with a master's instead. Oftentimes, people enter PhD programs without knowing what's expected. And while they're passionate about the subject, they find, justifiably, that the stress, the hours, and the foregone salary aren't worth it for them.

I would guess that many readers would be surprised to learn that PhD students earn a salary at all. Some of this salary is covered by university funds in the semesters that the graduate student teaches. However, most of it comes from government research grants. In either case, it's a small sum. I remember being delighted to find out, on a field trip to a chemistry lab at the University of Rochester in 1999, that the graduate students we met earned $17,000. Six years later, as a first-year graduate student at Harvard, I made $25,000. The figure at Rochester in my field in 2025 is $34,500. At UC San Diego, it is $40,000. At Stanford, an outlier by a mile, it is $56,000.

The quality of life one can enjoy on such a stipend depends significantly on the cost of living. While some universities offer subsidized housing to graduate students, this is not the norm, or at least not for the duration of the PhD. In any event, these subsidized housing rates also scale with the local housing market. The median annual rent for a one-bedroom apartment in Rochester is $15,600; in La Jolla, the upscale neighborhood where UC San Diego is located, it is $33,600. A PhD student at UC San Diego must thus incur debt to cover everyday expenses, at least if they hope to have a bit of privacy. The dire financial straits graduate students often face in expensive coastal cities have led to unionization and work stoppages, such as the strike that engulfed the University of California—including UC San Diego—in late 2022.

While scientific research amounts to just over 1 percent of the federal budget, these outlays are highly leveraged. The largest costs of research are borne not by the taxpayer but by the PhD students, who forgo both the higher salaries of their college classmates and the accrual of investment interest on any savings. Universities could always increase the salaries. However, funding agencies—who report to Congress—expect to get research results at a predictable rate. For example, if the NIH is used to awarding a grant at a rate of $100,000 per year, it expects to see results as satisfying as they have been in years past. If the cost of paying a graduate student suddenly doubles, then half as much research will get done, and half as many PhD graduates will be available to create value in the economy.

So why do college seniors clamor to get into PhD programs, despite the low pay and the significant risk that they will not finish? Many students, übernerds like me, really like school and research, and want to keep doing it. Also, there is a salary premium for PhD holders over BS and MS holders in the same industries. This seeming advantage evaporates when one considers that those higher earnings come much later in life. A net present value calculation reveals that PhD holders in biology barely break even by retirement age, while PhD holders in chemical engineering may not. Nevertheless, nonfinancial benefits—especially for those entering the professoriate—are significant.

Jobs for PhD-level scientists and engineers come with more professional agency, and often managerial roles. The social status of being called "doctor" or "professor" is significant, and high status leads to lower levels of stress over a lifetime. If a person finds that they truly love the intellectual freedom and creative outlet of research, the PhD is the way to go. One fact also not generally known to outsiders is that PhD students in STEM pay no tuition. It is not that PhD education accrues no costs. In fact, they are significant. The buildings, lab facilities, administrators, coursework, and their advisors' time are all expensive. Depending on the university, these costs of education are either absorbed by the campus budget or, more commonly, charged to federal grants.

Though a university is a nonprofit, it is still a business and must pay for its people and things. To do so, it sells degrees, produces research, and makes various investments. Some also provide medical services and entertainment such as sports. The university sees the PhD student as part student, part worker—an engine for both scholarly reputation and financial opportunity. Along with the students themselves, the university attracts research dollars and generates cachet—a kind of brand equity— for anyone associated with it. To accrue prestige, it needs to curate the highest-performing student researchers and postdoctoral scholars possible. If you're one of these students, it's a risk assessment. Depending on what advisor you're assigned to, you—as in you, the student—may have some level of control over your education, or your experience may differ wildly from that experienced by a classmate.

Which is to say, each PhD experience can be vastly different, and that makes these programs particularly hard to quantify. Nurturing, present, and emotionally supportive advisors—the kind who see their students through ups and downs and offer creative support—provide alternative experiences to what I had at Harvard. Which is not to say I would give up my experience of having studied under George Whitesides. I learned things from him and his group—selecting problems, scientific communication, research strategy, and personal independence—better than I could have in most other places. With a strong dose of luck, I ended up flourishing, but many do not.

In the natural sciences, a high percentage of students will "master out," or finish with a master's degree rather than move forward with a full PhD. At the top schools, master's degrees aren't offered in chemistry, biology, and physics at the time of admission. If you have a master's degree in chemistry from Harvard, it means that you "failed" to get a PhD. Finishing the PhD typically requires another four to six years. Given the low number of positions available for PhDs in the natural sciences in industry, many students stay longer in academia in postdoctoral—"postdoc"—positions. Most institutions have limits on the number of years someone can be classified as a postdoc. We have a pejorative term in academia, known, perhaps, only to us: the "permadoc," a PhD scientist who has been employed so long in a low-paid role in an academic research lab that the rules technically don't allow this individual to be called a postdoc any longer.

My own PhD took five years. In my first year, my stipend was covered by my work as a teaching assistant. The next four years were funded by federal grants from the National Science Foundation obtained competitively by my advisor, George Whitesides, who wrote the proposals that brought the funds to his lab. Whitesides had a success rate much higher than the national average. His ideas were innovative, and he was constantly coming up with new ideas. He used to quote the philosopher Schopenhauer: "Talent hits a target no one else can hit; genius hits a target no one else can see." Of course, he used this phrase at an award ceremony to introduce another scientist, but he might as well have been talking about himself.

Ideas require one set of skills, and execution requires another. A scientist who does experiments (as opposed to one who works only with theory) must have good hands. Laboratory work in the repetitive environments of chemistry and biology requires timing and dexterity. In physics and engineering, you often have to build your own equipment. Some of the most successful trainees with whom I have worked have been car mechanics, electricians, bicycle repair technicians, and military veterans.

In many ways, a lab worker is like a line cook. You rarely see this person from the front of the house, but anyone who's lived in biotech or deep-tech research knows exactly what's going on behind the doors. Every scientific task has its own tools, ingredients, and physical setup. The same way a kitchen lives or dies by mise en place (a French culinary term that means "everything in its place," referring to the organization of ingredients and preparedness of one's station), the lab runs on glassware, reagents, and rituals.

Let's take the mise en place of a chemist working in a university lab or the pharmaceutical industry. Almost all the work happens in uniform: a flame-retardant lab coat, safety glasses layered over prescription lenses, and nitrile gloves—usually blue or purple. Buy them on Amazon and they're dirt cheap; buy them from a lab supply catalog and you'll pay many times more.

Universities are a veritable playground of cool scientific equipment. Smaller items—say, a few hundred thousand dollars or less—are usually purchased with grant money applied for by individual professors. Larger items within this range are obtained using proposals written by teams of professors. In the grant proposal, researchers make arguments as to how the item is going to benefit a range of scientific activities. Items in excess of this range—large electron microscopes, spectroscopy tools, tools for making chips, and the specialized rooms they go in—are usually purchased using the university's capital budget. The capital budget, as opposed to the operating budget, is the same budget category that pays for buildings and renovations. There might be a campaign for philanthropic donations to underwrite such expenditures, but the bulk of the cost is often borrowed.

These items generate research dollars—for example, if they enable a fancy new type of experiment—and also user fees, which can be used to service the debt incurred by the loan. User fees are paid from faculty research grants, every time one of their students logs in to the instrument, and also by local companies that need access to tools they don't have. Small and even larger companies can spend a lot of money using university facilities for basic research and development. While internal users (students) pay reduced rates for access to the equipment, industrial customers pay full price.

For smaller pieces of equipment, say the cost of a Mercedes sedan or less, the upkeep and maintenance often falls to students and postdocs. Anyone who has spent enough time in a university lab can be expected to have expert-level knowledge in the workings of fine instruments and even moderately heavy machinery. At Harvard, I used to spend days monkeying with a huge high-powered instrument called an electron-beam evaporator. This instrument, designed to deposit thin films of metal on computer chips, did so by melting Reese's cup–sized blobs of gold and platinum at thousands of degrees Celsius and under intense vacuum. One time, after cleaning and reassembling the chamber, I failed to secure the water coolant lines and flooded the instrument and the entire lab.

Setting aside these sources of exasperation, the research lab is full of hazards to a person's physical safety, and they come in both acute and chronic varieties. Pressurized equipment, explosive reagents, and flammable solvents. Just to maximize the damage to human flesh, we use glass for every reaction, distillation, and analysis. Most of this glass is not safety glass. In fact, it seems to be designed to be as sharp as possible when dropped or blown up, and we've all been bloodied by its shards, even if we've never been in the blast radius of a full-on explosion. My lab goes through thousands of hypodermic needles and syringes each year too. We use them for everything: to measure liquids and to pierce rubber septa sealing sensitive reagents. A close cousin of the needle is the long, wirelike cannula—beveled needles at both ends—used to transfer air-flammable liquids from one vessel to another. Apart from the poking hazard, the cannula carries the risk of the accidental detachment of one end and the release of air-sensitive liquid into the air. Some of these compounds spontaneously combust in air.

Among the more insidious, long-term hazards are the carcinogens, which often arrive in the form of highly reactive molecules that modify your DNA; teratogens, which is why women are advised to stay away from many types of labs during pregnancy; dust from drying agents, including ubiquitous microfine silica gel; acids and bases (recall my own trip to the emergency room); and radioactive isotopes, which are especially used in cell and molecular biology. Lethal burns aside, all of us have ended up with mysterious rashes from accidental contact with these substances.

The goal of a researcher and the campus department of environmental health and safety is not to eliminate exposure but to dilute it. An N95 mask still lets in 5 percent of particles and contaminants, and no amount of air changes per hour in a lab working with organic compounds will prevent at least a million of those volatile demons from going into your lungs, blood, and brain. You're worried about drinking microplastics in your tap water? I got you beat. The trade publication of the American Chemical Society, *Chemical and Engineering News*, publishes weekly obituaries of its members. The good news is that very few of us seem to perish from heart disease and other illnesses of old age.

In some professions, a person takes on personal risk in exchange for increased compensation. It is why butchers, who work in a refrigerator all day using cutting tools designed to slice flesh and bone, make more money than bakers. Similarly, if you join a startup company that doesn't have the money to pay you a high salary, you expect to get stock options, which might make you rich if the company is successful. A graduate student, on the other hand, takes on risks to physical, psychological, and financial security. As a graduate student, I maintained my paycheck-to-paycheck lifestyle by withdrawing $120 per week in cash from the ATM. That was all the money I got to spend on anything besides rent and utilities.

Once, when I visited the bank to resolve a fraudulent charge, the agent stepped away from her computer. I peeked at the screen and saw a box containing the words "Retention Score: Low." The bank made the cold calculation that I didn't have much money and that I evidently wouldn't have much in the future. Partway through graduate school, I sold my beloved trombone to make rent, somehow not believing my financial situation would ever improve. Now I would spend ten times what I sold it for to get it back. The promise made to the PhD student to compensate for the risk—in lieu of stock options or higher pay—is a lifetime of nonpecuniary benefits: the prestige of being addressed as "Dr." in your junk mail and, undoubtedly, the sense of accomplishment of beating the final boss of the game of academia.

Defending your PhD thesis really does feel like a boss fight in a video game or the final scene in a police drama. It takes place in an austere, courtroom-like setting. Your adversaries are formidable: a panel of professors who are experts in your area of research, if not your specific project. The battle takes place years after your last formal course, probably in the spring of your first year of graduate school, though you are still responsible for remembering the content of those courses. There is no formal duration for a PhD. You defend when your advisor either suggests or acquiesces to a date. The date can be set if you solve a scientific problem, you publish some typical number of papers, you get a job offer, the lab is out of money and needs to shed payroll, or through some combination of these factors.

In chemistry and materials science—and this is different in every field—you need something like three first-author papers or more to write a thesis and defend. In the physical sciences and engineering, the "writing" of a thesis is a joke. Whereas a PhD candidate in English takes years to write a tome that becomes their first book, a materials scientist simply takes their papers that are already published, does a bit of formatting, and sends them in to the book binder. No one *ever* reads the PhD thesis of a physical scientist or engineer, at least not one in a subfield served by many journals. Any useful data has already been published. There are other fields, typically more fundamental ones like cell biology, physical chemistry, and astrophysics, in which the opportunities to publish are fewer. Materials science and many fields of engineering are served by more journals because they are inherently combinatorial. A person can make a new material, learn how it works, and then publish a paper. In biology, once you learn how a certain type of virus infects a cell, that's it—a few papers, but no more.

During my PhD I published five first-author papers. Whitesides agreed to let me defend at the end of my fifth year, in June 2010. I assembled my committee: Whitesides; Hongkun Park, the Mark Hyman Jr. Professor of Chemistry, who terrified me; and Joanna Aizenberg, the Amy Smith Berylson Professor of Materials Science and a professor of chemistry and chemical biology.

I had chosen Park, specifically, for the challenge. He had a reputation for being tough, and if I was going to have a PhD from Harvard, I wanted to be put through the wringer. I had gotten married in April—more on that later—and I had attended my wife's PhD defense on the previous day. On my big day we found ourselves in a beautiful, classic, oak-paneled seminar room in historic Pierce Hall, truly emblematic of the Ivy League experience, exactly the kind of rooms you picture when you imagine Harvard. The night before, I snuck into the room and practiced my entire defense. The day of the defense, Whitesides, Park, and Aizenberg arrived, followed by the students from the lab and other Harvard community members. The candidate is expected to deliver a public lecture, followed by a public Q&A session. Then the public audience leaves and the committee grills the candidate in private. At least this is how it's supposed to go. My public talk went well. Afterward, members of the audience asked a few questions, with a few softballs from the committee. Aizenberg asked something easy, meant to satisfy her curiosity rather than being a gotcha, and then Whitesides asked a single question.

"So, Darren, you've been working in nanotechnology for five years now," he said. "What's your assessment?"

My *assessment?* The question was so vague as to be comical. I took a deep breath. "Nanotechnology has been around a long time," I said. I then repeated a line of thought I had heard from Whitesides himself. It came out like word salad. "Nanotechnology has been around, but it hasn't always been called that. On your glasses, for instance. You've got nanostructured thin films, which prevent scratches. And DVDs. They also have nanoscale features on them, which the laser scans. The trick to nanoscale engineering is repurposing existing tools, such as microscopes and laser scanners, so that they can be used to make nanostructures, rather than just image them."

Whitesides looked pleased enough. He said, "OK, at this point the audience can leave. Darren, you can leave too."

I looked at Park as though he were a predator, lying in wait. This man, who had scared me so completely, had asked nothing at all. I looked around. The space was emptying.

Whitesides looked at me, and then at the committee. "Unless we're actually going to disagree about the outcome of the exam," he added.

Park just sat there, in tacit agreement, either in deference to Whitesides or because he was glad that the exam had wrapped up sooner than he had expected.

I was stunned. There was no technical question. No merit-based assessment of my work. My thesis wasn't exactly a basic topic for beginner scientists. I had expected the committee to insist on asking a *real* question. What's the free energy of that reaction? Or anything—any single thing—that reached back into the *science* of what I had been doing for the last five years.

I'll admit I was rattled—not by the exam itself, or even by my inquisitors, but by how *easy* it had been. I had steeled myself, had prepared myself to be raked over the coals. To endure the quintessential rite of passage of the PhD experience. It was also a blessing—a lucky resolution—deus, nay, Whitesides ex machina. A PhD defense was, apparently, like everything else about a PhD program, a simple rolling of the dice. *You get what you get, and you don't get upset.* Good luck, bad luck— either way, it came down to luck, and any other student would have agreed: On that day, I got very lucky, indeed.

Two minutes later, my committee walked out, shook my sweaty hand, and extended their congratulations. I had completed my driver's test without ever having to parallel park (which also happened to me). It was over. I had my PhD.

That one-hour exam—an unusually short one for a PhD defense— concluded my stint at Harvard. The feeling was one not of exhilaration but of release, an immediate letting go of the stress of graduate school at Harvard that quite literally sent me to the hospital—twice (once for the acid in the face and once for a cardiovascular crisis; see next chapter). Oh, what a ride it was. I entered the school as a young and aspiring scientist, naive to the academic research world. When I left half a decade later, I was wiser. I won't say that becoming a PhD jaded me, but it certainly gave me a deeper understanding of the hurdles that face candidates as they sacrifice earnings, outside interests, and mating opportunities in the service of science.

Chapter 6

By the time I graduated from Harvard, I had already met, dated, and married my wife, Dina, a fellow scientist who holds a PhD in virology from Harvard. Her name is homophonous with that of my sister Deena, a source of confusion my family still hasn't quite figured out how to negotiate. My relationship with Dina is foundational, so I'll backtrack a few years.

My romantic success, I should note here, had been nonexistent before I met Dina. I never got out much, so by the time I got to Harvard, I was considering alternative ways of meeting a partner. I spent a long time deliberating the merits of online dating before I finally caved and created a Match.com profile. When I met Dina, I had been on Match.com for nearly a year and was going out on one or two dates per month. I think a lot about my percent yield—even dating can be framed as a chemical reaction—and I would send out about ten emails and receive one back with interest. I treated relationships the way any good nerd would: like a science project.

For each of the ten or so potential partners I met on Match, we didn't share enough mutual interests to continue. It's common for scientists to date within their own profession: We see the same people day in and day out, doing repetitive tasks for much longer than the standard work week. Also, we speak a language in our professional worlds that is incomprehensible to outsiders.

One mistake I made in my scientist's approach to dating was thinking that I needed to go to froufrou restaurants on first dates to demonstrate . . . what, exactly? Advice to a fictional younger brother would be, "Dude, you're a grad student. She *knows* you're poor!" So I eventually decided I would only go to places where I felt comfortable and that wouldn't break the bank.

The first date with Dina almost didn't happen, since on my profile, where it asked how many drinks I imbibed per week, I put "seven to fourteen." I figured that was on the low end for grad students in organic chemistry, not known as a teetotaling culture. I learned later that she initially passed on my profile because she thought I was a low-key alcoholic. Rather, I thought she'd bristle at a different one of my responses. For body type I had selected "athletic and toned" from the drop-down list. This choice was only a notch above "about average," which was closer to the truth, given what I had selected for my drinking habits; on the other hand, I was running five times a week. In terms of my health, I was driving with the brakes on.

For our first date I suggested Cambridge Common, a restaurant and beer bar on Massachusetts Avenue between Harvard and Porter Squares. Because I had no experience with romantic relationships, I was convinced of this one's imminent failure. I had always dedicated my time and attention to grades and academics. I also suffered from crippling anxiety, and the physiological effects of said anxiety, such as excessive sweating. This was before I knew anything about the magical effects of SSRIs and beta blockers. My anxiety was magnified in the framework of dating. How would a woman ever *want* to hold my sweaty hand? To say nothing of someone one day wanting to see me naked. The thought defied reason. During my adolescence, *Seinfeld* was hugely popular, and reruns played constantly on our local Fox affiliate. When I considered my anxiety and social paralysis, I used to recall the line, spoken without prompting by George Costanza: "You know, I've been thinking. I cannot envision any circumstance in which I'll ever have the opportunity to have sex again." But in my head, the line played without the "again."

By eighth grade, I had already resigned myself to a single life. *I'm just going to join the Navy and be on a submarine and be underwater for months,* I told myself. It would be fun. I'd get to visit the deep unknown and fire torpedoes at slow-moving enemies, just like serving on the Starship *Enterprise*. If I had any real system of belief about God, I would have been a priest as a social cover for my paralysis and fear. In high school my fellow band geek Adam and I made ourselves the sole members of a fictitious club we called Virgins Forever. To be a member didn't require some dour misogynistic identity held by modern-day incels. Rather, one simply had to be (1) a virgin and (2) amused by the government-mandated programs in school that promoted abstinence.

On our first date, I learned that Dina came from a culturally Muslim family: Her mother was born in Baghdad; her father, in Beirut, to Iraqi parents. She, in fact, grew up on an American compound in Saudi Arabia, as her father was a staff scientist at the King Fahd University of Petroleum and Minerals.

After our initial date at Cambridge Common, we went to a couple of other local favorites, including the live music venue Harper's Ferry, one of the watering holes where Aerosmith played early in their career. Another early date was an all-day Saturday event where we hit up the Arnold Arboretum, Boston Beer Works, and the North End. When we stayed in, we drank so much beer and ate so much homemade pizza that we each put on ten pounds. Dina was a member of the Harvard cycling team, and I was a runner, so we weren't overly concerned about the extra calories.

I was living with a high school friend in a basement apartment near Harvard Square, so I spent most of my time at Dina's. She had a ten-year-old cat, Mera, a tortoiseshell calico we only ever called Nunu for some reason. For the next ten years, until she died at twenty, Nunu would become as much my cat as Dina's. When Dina became a house tutor, kind of like an RA, during my second year at Harvard—our first full year together—we ate with the Harvard undergraduates at their dining hall.

Dina and I shared many interests, chief among them being music. Her favorite band is Metallica, and mine is Tool, which was her second favorite. We stayed up late many nights listening to grunge albums on my 1970s Pioneer stereo system, especially the *Jar of Flies* EP by Alice in Chains and the single album from the short-lived supergroup Temple of the Dog, featuring Soundgarden's Chris Cornell on vocals backed by Pearl Jam.

We also enjoyed hiking—in the White Mountains of New Hampshire—and travel. We traveled locally, on weekends, to several places in New England the first summer we were together. Neither of us had a car, so we would pick up a Zipcar and drive off. Our first major trip together was to Las Vegas to visit Andrea, who had, by then, moved to Sin City with Ronnie. I got the worst food poisoning I had ever experienced from the buffet at the Mirage, likely brought on by pink crab legs infested with norovirus. While I recovered in the back seat of the car, Andrea drove Dina and me to the Grand Canyon, where it snowed.

Some travel I did on my own. In 2008 I had applied to a competitive program through the National Science Foundation for the winter school in India, to take place the first week of December. On November 26, a series of twelve coordinated Islamist terrorist attacks took place in Mumbai, a sustained, four-day tragedy that claimed the lives of more than 160 people. India had initiated extra security protocols. Suddenly, the idea of traveling seemed illogical at best and dangerous at worst.

My anxiety started to spike. I experienced something referred to as a premature ventricular contraction, where the heart skips a beat. It was happening over and over again, maybe fifty times a day. Dina had started living with me by then. One morning I was convinced that my heart had stopped beating altogether. I was sent to Brigham and Women's Hospital, part of Harvard Medical School, for a cardiac ultrasound, a treadmill stress test, and a cardiac MRI. I met with a cardiologist and was released and told to wear a Holter monitor for a full day, and, after some discouraging findings, a full month.

During the month my heart was under 24/7 surveillance, I had hundreds of arrhythmias and even a few ventricular tachycardia events, where the heart reaches 180 beats per minute. Beyond those erratic events, though, my heart was normal. They believed, at first, that I suffered from an autoimmune disease, but the conclusion was the opposite. This was an idiopathic reaction, unlikely to repeat itself.

Cleared of any danger of life-ending heart disease, I felt freer to travel, to see the world. I ended up going on the trip to India, and it was life-changing, if for no other reason than it exposed me to just how flavorful food could be. Dina and I had other bucket-list destinations on our agenda. We began to make big travel plans. We wanted to see Peru. Borneo. Sicily. We got married and defended our PhDs. We moved to California to attend Stanford for our postdoctoral work.

Dina and I had just finished our postdocs at Stanford in 2012 when we decided to take a huge, transformative trip together. We planned a trip with a small group tour that was meant to be one of our momentous moments together on the road. Things didn't entirely work out the way we had planned.

Dina had always wanted to travel to Machu Picchu and had an enduring love for the place, ever since she was introduced to it as a child. She had first learned of it in the book *Prisoners of the Sun*, from *The Adventures of Tintin*, which describes Machu Picchu and the Sacred Valley. Her interest was piqued. To me, Peru sounded great. We had done our fair share of low-altitude hiking throughout the Santa Cruz Mountains, and a little into the Sierra Nevada too. The Santa Cruz Mountains level off just below four thousand feet. The Sierras are higher, with the taller peaks reaching elevations above fourteen thousand feet, though we never went up that high. I wasn't concerned about the fitness level required. By then, I was an avid runner and cyclist. There was, however, one small genetic factor that I should have taken more seriously.

A few months earlier, my sister Deena had alerted me to the possibility that I might carry the sickle cell trait. During routine testing at an annual ob-gyn appointment, she had had a blood test that had determined that she carried it. Our parents had known we had the trait, but they forgot about it, since it's supposed to be asymptomatic. While my own sickle cell trait status was buried in some ancient health record in Rochester, I discovered it for myself from a 23andMe DNA test, which came back with the genetic marker for the trait, most common among individuals with African ancestry, but also prevalent among Sicilian-Americans. Deena left the doctor's office with an information pamphlet from the 1970s, photocopied thousands of times, titled *Living with Sickle Cell Trait.* The grainy cartoon silhouettes of the people on the front page had afros, betraying the stereotype that only Black Americans carried the gene.

People with the sickle cell trait carry only one copy of the sickle cell abnormal hemoglobin gene and *usually* do not have disease symptoms. Carrying two copies leads to full-blown sickle cell anemia. In most people, red blood cells are round and flexible, allowing them free passage through the blood vessels. Sickle cell disease causes some of the red blood cells to take the shape of a sickle, and these cells can stick together and block the flow of blood.

Altitude can exacerbate the symptoms of sickle cell disease. At high altitude, where the air has less oxygen, the body is more prone to producing these misshapen cells. When the cells become sticky, they can clog free passage between organs, and certain organs—the spleen, say— can be more prone than others to receiving the burden of these blockages.

I hadn't thought much about sickle cell disease until that point, to be honest, apart from the fact that the diagnosis was unusual. I didn't know anything about the prevalence of sickle cell trait among Sicilian Americans. I soon learned that 13 percent of people living in Sicily have this trait because it is predominant in areas that are or have historically been susceptible to malaria, including the Mediterranean, the Middle East, and North Africa.

The aftermath of my genetic revelation had not, however, inspired in me any real fear. My prevailing sentiment, to the contrary, was *What are the odds that anything is going to happen?* It seemed, from the literature, to be extremely rare.

Hikers of the Inca Trail typically first arrive in Lima and then connect to Cusco in preparation for the remainder of the trip. Cusco is over eleven thousand feet above sea level, and visitors to the Sacred Valley use it as a place to help acclimatize. We were spending a few days in Cusco, getting used to the thin air. On our first night I got up to go to the bathroom and collapsed on the floor. I'd never fainted in my life. I don't know when I regained consciousness or how long I was out for, just that something overtook me that was deeply unfamiliar.

I dismissed this incident as boring old altitude sickness. I was in denial—everyone suffers a little on the first night at high altitude. People complain of headaches, of lightheadedness, of the inability to breathe when carrying out simple tasks. None of these symptoms had come for me. I had only fainted! What was the big deal?

We started the multi-hour drive to the staging area. Throughout this journey, I identified a pain in my side, steadily becoming more intense. Just stomach pain, I told myself, because of fluctuations in oxygen in the gut at high altitude. I was also eating unfamiliar food, including grilled alpaca, which I ate despite being a vegetarian at home.

It just *happened* that the increasing pain was localized right where the spleen was. I was taking acetazolamide, the standard high-altitude sickness medication, to ward off headaches. I felt protected—at least against the effects of traditional high-altitude sickness. Yet I couldn't lie on my back because it put too much strain on my torso. I could barely eat or drink; any expansion in my stomach against my ribcage was excruciating.

We reached Dead Woman's Pass, the highest point on the Inca Trail to Machu Picchu. The altitude: a staggering 13,829 feet. Named for a vista formed by the surrounding mountains that resembles the silhouette of a woman, the pass is the most challenging section of the ascent. From there, the hike dips down into the Sacred Valley toward Machu Picchu, itself a mere 7,972 feet. Dead Woman's Pass was almost unbearable, and we spent an entire night near the summit of the trail. Descending toward Machu Picchu, I could feel the pain start to release. Our hike complete, I decided to stop taking the acetazolamide. We drove back to Cusco, where I had been fine. Cusco is still 11,152 feet, and at dinner, I was attacked by vicious pain.

"We need to leave," I told Dina. I was beginning to have trouble breathing. We stopped at a pharmacy, our best hope for anything resembling urgent care. They hooked me up to emergency oxygen, piping it in through my nose.

All night I tossed and turned, unable to breathe. It was as if I couldn't engage the autonomic part of my brain stem; each breath required great effort. I was afraid that I might fall asleep and die because my brain was no longer able to inflate my lungs. I couldn't sit comfortably in a chair. I couldn't lie down. I didn't want to sleep, so I took a pile of pillows and found one position where, if I hugged them and hunched over, I could get through the night without triggering excruciating pain. We were scheduled to spend the next five days in the Amazon rainforest but threw those plans in the bin, as medical care would have been even more remote than it was in Cusco.

The next morning we boarded an emergency flight back to Lima. We were close to the equator, but I felt so cold that I wore all the clothes I brought with me, and my face was still ghostly white. I had lost eight pounds in four days. Shuffling to the departure lounge, I prayed, an atheist in a foxhole: *Please don't let this plane increase in altitude very much more before it reaches Lima.*

Biologically, I know a little more now about what was happening. With the return to lower altitude—about five hundred feet above sea level—my red blood cells had stopped sickling, meaning that the spleen got a break. A splenic infarction, which I was experiencing, is caused by the blockage of red blood cells, and it can cause parts of the organ tissue to die. We went to Lima's emergency clinic, where, unlike in the United States, they don't even ask for your insurance. You simply swipe your credit card to pay for the tests, the way you would after ordering a coffee at Starbucks. It cost forty-six dollars to see the doctor, for the splenic ultrasound, and for a complete blood and metabolite panel. If I were uninsured in my home country, I would have received a bill for thousands.

I received a prescription for Celebrex, a nonsteroidal anti-inflammatory drug. I was disappointed I didn't get something stronger. I was cleared to travel within a few days, assuming my pain level was low. I was advised to see a doctor when I returned home. We spent the next few days in Lima, where I was eventually ambulatory enough to visit a few tourist sites, including the Convento de San Francisco Ossuary, a catacomb that holds the bones of an estimated seventy thousand people.

Then we went home. My primary care doctor at Stanford admonished me for going in the face of the diagnosis of sickle cell trait.

Despite how my genetics rebelled against me and I nearly died, I wouldn't trade the experience in Peru for anything. For one thing, the Peruvian Andes is the most beautiful place I have ever been. Thankfully, I'm unable to resurrect the pain, but I can remember exactly how I felt when I saw the sun rise at Machu Picchu. More importantly, my relationship with Dina was inexorably deepened in a way accessible only by shared trauma. She had good reason to think that I might die and fought tirelessly on my behalf to get me to medical care and, ultimately, home.

Two years later, on a trip to Malaysian Borneo, I was, unfortunately, able to return exactly this favor. Just as Dina's interest in Peru was spurred by a childhood encounter with Tintin, my dream trip to the island of Borneo was sparked by an oft-borrowed VHS tape produced by National Geographic, owned by the Parma library. The program, *Creatures of the Mangrove* (1986), pictured on its cover a photo of a proboscis monkey with a huge floppy nose. The mangroves, the rainforest, and the mountains of the landscape inspired me completely. Decades later I finally had the means to travel there. We saved up our money and booked a trip.

Borneo, the third-largest island in the world, comprises parts of two countries—Indonesia and Malaysia—and the whole of a third: Brunei. While we were walking on the sidewalk in Kuching on our first evening on the island, Dina was yanked into traffic and dragged more than fifty yards by a thief on a motorcycle who was trying to steal her purse. A seasoned traveler, she would normally have kept her valuables under her clothes, but this was just going to be a short trip away from the hotel. The thief got away with her phone, money, and passport. More critically, she was bleeding—skinned well into the subcutaneous tissue—from both ankles, knees, hips, elbows, palms, and shoulders.

Witnessing the event, which had stopped traffic and drawn a crowd, a wonderful Malaysian family drove us, along with their kids in the backseat, to the police station. There, I washed Dina's wounds in the bathroom, which, along with being intrinsically grimy, had no soap. The police report proved invaluable to get a temporary passport when we eventually made it to the US embassy in Kuala Lumpur (hundreds of miles away across the South China Sea). From the police station, we were able to contact our tour guide, who borrowed his cousin's car to take us to the hospital. There, the veiled nurses cleaned her wounds and the physician determined that there were no broken bones. They loaded us with antiseptic and gauze pads and sent us on our way. Unlike Peru, which charged an exorbitant forty-six dollars for a visit to the ER, emergency medical care in Malaysia was free.

During this trip, cut short like our trip to Peru, Dina was stoic. She has a quality I sometimes describe as "digital despair"—on or off. The level of intense despair in the "on" state was the same for life-and-death scenarios as it was for silly pet peeves like the dishwasher not being loaded in a certain way. But when finding emergency oxygen for your suffocating husband or walking in the rainforest covered in bandages or making emergency travel plans, it is a godsend. These ordeals have deepened our love and connection and have made other challenges child's play by comparison.

Chapter 7

After getting a PhD, the next step in becoming a professor is to complete a stint as a postdoctoral scholar. The postdoc for a scientist is analogous to a residency for a medical doctor. You have your terminal degree but need to show a greater degree of independence. Postdocs are required for nearly all professorships and for many industrial research positions, especially in the pharmaceutical industry.

I had my sights set on doing my postdoc with Zhenan Bao, who was then a rising early-career professor but is now approaching the league of George Whitesides in terms of productivity and influence as a scientist. In 2004 she had taken a position at Stanford after a successful nine-year position at Bell Laboratories in New Jersey.

You have to think years ahead when applying for your postdoc and make your own introductions to prospective mentors. Whitesides usually didn't go out and make these connections for me, although some PhD advisors might have assumed that role. His name, however, carried a lot of weight, so a PhD under him could offer a lot of choice when it came to postdoctoral advisors. You didn't *need* him to go to bat for you. You could just drop his name, the H-bomb of scientific advisors.

Bao came to Harvard to give a talk as part of our normal seminar series. In academia, every department has a weekly seminar series during which an outside professor comes in to meet with other professors. There might be, say, a luncheon, in which students are included, or a coffee hour. Normally, however, the faculty co-opt the visitor's time. These people, who have so much to offer and who students see as shining beacons within their field, are never really all that accessible to them. Nevertheless, I shoehorned my way onto Bao's schedule by emailing a member of the front office staff with whom I happened to be friendly.

Bao had one open slot on her packed itinerary. If an additional faculty member had wanted to meet with her, I would have been bumped. Still, my persistence paid off.

It was the fall of 2008, just before the bombings in Mumbai. I told Bao about the jelly roll paper, along with some of the other ideas quite literally cooking in the lab. "I hope when you apply for a postdoc that you consider my lab," she told me. I was floored. It was what I had been hoping for, of course—an invitation. It was the reason I had made the effort. Still, here she was, a rising-star scientist, soon to be a supernova. And she was inviting me into her lab.

In reality, I knew that Bao was still a person—a very real person, just like me, who had paid her dues and done her research and climbed the rungs until she reached the top of the ladder. That reality did not connect with all of Bao's outsized accomplishments. For example, at Bell Labs she had done some of the pioneering work on the organic field-effect transistor, a microscale electronic switch that has proved to be incredibly useful in research and which became the basis of several other technologies. She was working at the cutting edge of a field that touched on solar cells, lighting sources, and biosensors, and doing so in a place—Silicon Valley—that seemed poised to capitalize on inventions coming out of her lab. Even I had trouble believing that it was actually coming to pass.

Eight months later I found myself in San Francisco for the annual meeting of the Materials Research Society. I took the Caltrain to Stanford and attended a job interview with Bao and her already large team of graduate students and postdocs. Not long thereafter, I received an offer. My start date wasn't until the summer of 2010, so this was well in advance—more than a year out.

I still had to apply for funds, because even though Bao was committed to me, it is considered a sign of good faith to at least try to obtain your own funding. Most postdoc fellowships in the US are open only to trainees within the biomedical sciences. There was one, though, the intelligence community postdoctoral fellowship, connected to the

CIA and NSA, which funded prospective postdocs in the physical sciences. I applied for the fellowship, leaning into the topic of skin-conformable wearable sensors for covert operations.

My proposal was selected for funding, and I was able to anoint myself with the honorific of "Intelligence Community Postdoctoral Fellow." The award paid about 20 percent more than what Stanford was paying postdocs, which was useful given the exorbitant cost of living in Silicon Valley. The fellowship also came with some money for materials and annual travel to visit CIA headquarters. (As cool as that might sound, they only let you see the lobby, the museum, the gift shop, and the restroom.) Most importantly, since I was self-funded, Bao wouldn't have to pay me out of her own grants, and she was grateful to be able to free up her funds. I could go back to defending my PhD, secure in the knowledge that I had somewhere to go after Harvard.

There had always been, to me, something magical about California. Dina and I were obsessed with the late chef and Food Network star Michael Chiarello, who had cultivated an of-the-moment Napa style. When I was applying to graduate schools, I had been accepted to four different places in California—plus Harvard, of course—and in my imagination I had considered how nice it would be to escape the cold and the snow, to go to California and to be on permanent vacation. Nevertheless, I ended up at Harvard, because no one in organic chemistry says no to Harvard.

While packing our belongings, I listened to "Go West" by the Pet Shop Boys, sent to us via Andrea on Facebook, pumping myself up for the move. Palm trees. Beaches. A long slip of coastline. Great weather. Surfers. The West Coast felt full of possibilities, all of which lay before me. Anyway, what did we have to lose? We were young, we were free, we had all the time in the world to make mistakes or not make mistakes, to grab life and do whatever we wanted with it. We thus gathered up Nunu and moved to Palo Alto.

At first it was a big adjustment. Although Whitesides and Bao are two of the most famous material scientists alive, their styles were nothing alike, from project selection to mentoring. Whitesides's approach was to do something completely new—to make people furrow their brow and say, "What on earth are you doing, and why?" Eleanor Hutcheson's worm clamp was a case in point. He would say that the scientific enterprise rests on a three-legged stool, and two of the legs are the familiar research and development, but the third is invention. He focused on getting from zero to one on a new idea, as opposed to going from two to ten (research) or eleven to one hundred (development). Whitesides regarded all the fun to be in the creation of something from nothing. He used to say that working in a field that you invented (one with no competitors, at least initially) was a form of risk mitigation. If you were the only person doing something, you didn't have to read the literature—since if it was sufficiently new, there would be no literature—and you didn't have to do it very well because there were no competitors.

Of course, he relayed this conceit to the large crowds at scientific conferences as a joke, but there was a grain of truth to it. If a scientist works in a crowded field, there is a risk of being outcompeted while pursuing the same outcome. The contribution of a second- or third-place finisher is easily forgotten. An "also-ran" could be said to waste precious resources that could have been more strategically directed elsewhere. Students working on less successful approaches to popular problems are less likely to get high-profile jobs in academia and industry. Graduate students and postdocs who had success in the Whitesides lab all bought into the approach: Stay original or risk making somebody else more famous.

Bao, on the other hand, attracted great grad students and postdocs, and her work was closely connected to industry. Samsung. Apple. HP. These companies sent people to her to train them, and they would give her lab money as compensation. There was, then, an expectation that the people who went to her lab would be trained with applicable skills that could be used within the real world. She also attracted an enormous amount of grant funding and participated in several government–private research centers on campus. In addition to sending students to the flourishing hardware (now called deep tech) sector in Silicon Valley, Bao also sent many people to academic positions, myself among them. If I can make one comparison between the "typical" trainees of Whitesides and Bao, I would say that the graduates of Bao's laboratory tended toward a higher technology readiness level, or TRL, by the standards of academic science. Her motto, if she had one, would be more practical: We're going to embrace competition with other research groups on longstanding problems but do them better than anybody in the world.

I worked in two areas within Bao's lab. The first area was one I will claim to have invented: "rubber" solar cells. A solar panel that is on a roof, for instance, is subject to a lot of wind force and hail. Maybe tree branches fall on it. Maybe disaster relief forces like the Red Cross put up tarps and tents, and maybe you need them to generate electricity. Within this niche application, there's a cool scientific problem: The materials that are good for electronics tend to be rigid and brittle. If you can make these devices more flexible and even stretchable—more like rubber— they can be jabbed, dropped, and bent without breaking. This capability also offers other tie-ins to biomedical research; for example, these flexible devices can be put in the body and on the skin, where they will conform to biological structures and interact with body parts and internal organs without damaging the tissue.

My postdoc years were tough, not because of my time in Bao's lab but because it was a time of great personal transition. In this case, as would be repeated two more times in our marriage, we moved because of where *I* found a job, the "leading spouse." It was the location of my job that dictated where we were going to live, and that would constrain Dina's choices. Dina's PhD took six years compared to my five years, not because she worked less hard than me but because her PhD was in fundamental mechanistic science—virology, in her case. In fields like this, research progresses more slowly and there are not as many papers to be had as there are in engineering. She graduated with one first-author paper within her PhD years to my five. I had been in a lab with forty people; synergy is a big deal in a big lab, as you work and publish many papers with lots of authors. In her PhD lab, it was only her and a senior postdoc.

While Dina is a brilliant scientist, landing a postdoc position at Stanford or UC San Francisco was a struggle. She had applied to four labs, was declined from three, and was accepted to the lab of Professor Harry Greenberg, a gastroenterologist at Stanford Medical School. Greenberg's lab was located at the VA hospital in Palo Alto, a long trek from the Stanford campus and thus nowhere near the fun academic community that she was seeking. I would ride my bike to Stanford, and she would drive to the VA hospital to spend her days in a strange environment. It was isolating and depressing. To decompress, we went on long bike rides in the redwoods. She expressed her dissatisfaction with her position, and I didn't blame her. We had both wanted to move to California. We both had jobs, and not everyone could say that. And, at the time, you could live in the Bay Area on two postdoc salaries, rent a seven-hundred-square-foot apartment, and have a pretty nice life. Still, looking back, those days were among the hardest for us.

My postdoctoral fellowship in the Bao lab got off to a fast start. It helped that I had been in such an unstructured lab at Harvard. Whitesides didn't hold our hands, and if you couldn't do at least the doggie paddle, you drowned. The independence allowed me to work at a fast clip at Stanford. I would send Bao complete drafts of papers that she barely had to comment on. I wrote five first-author papers during my two-year postdoc, which equaled my total over a five-year PhD.

When I think about my time at Stanford—and, more wholly, about the type of person who is drawn to science—I often think of Seong-Ho Choi, one of my labmates during my first semester. Choi was a fellow postdoc. One day he didn't show up for work. A few more days passed. Choi still had not appeared. Someone initiated a wellness call, and the police went to his apartment. Choi had died.

He had suffered from an infection—one for which he had never received treatment. He had tried to rest it off, but the infection was aggressive. Too aggressive, it turned out. Right after his death, my labmates and I were invited to a memorial service hosted by Choi's family. Some came all the way from Korea. We were meant to hold his funeral service in a church in the evening. Choi had been part of the Korean Christian community. The service was set to begin at seven, and at around four that afternoon, I went into Bao's office.

"I have this data that I'm really excited about," I told her. I was enthusiastic. Just thinking about how excited I was, all these years later, embarrasses me.

"That's great," she told me. "Keep it up."

Later that evening she delivered his eulogy. Bao was distraught. I had never seen her this way: tender, human, and sad. She was a person with deep emotions. Somehow, in the midst of my own needs, I had missed her needs. I had missed the fact that she was a person, too, and not just a supernova.

It hadn't occurred to me how self-centered I had been—or how self-centered my work had made me. Our labmate had just died. There in the pew, I found myself wrestling with that old, familiar demon: shame. How shameful that I had bothered Bao about some lab finding only hours before she was to memorialize a life. A human life. A life that could have been any one of ours. I vowed, from then on, to be present for others, to show up when they go through loss, even if it feels unnatural.

Things were still going well at Stanford. But I didn't know whether or not I wanted to be a professor. I was hedging, and intentionally so. I didn't know what kind of CV I would have after my work at Harvard and Stanford; it looks good on paper, but almost every professor who gets a job at a top-five research university has at least one of these schools on their résumé, so I was barely differentiating myself.

To try to find some clarity, I stepped momentarily outside of science. Stanford Business School had an offering called the Program for Innovation and Entrepreneurship, now called Stanford Ignite. It was a night and weekend program that lasted a whole semester: fifteen weeks, two nights a week, and one full Saturday a month. Each week was taught by a different member of the Stanford business faculty who gave us the "gold coins" from their areas of specialization. So, we got compressed versions of the core courses in the MBA curriculum: accounting, finance, marketing, operations, strategy, startup formation, management and soft skills, and executive compensation. In the background we worked with a team of fellow scientists, doctors, and engineers to develop an idea from lab to market. While these projects were intended to be mock versions of how ideas turn into commercial products, some of the teams actually were funded and became successful businesses.

Dina and I both took the program. It was competitive to get in, but if you did, you were offered a massive discount as a Stanford student. It cost us only $500 to complete the entire program—the best money I've ever spent. People from industry who came in paid the full price: $10,000. Doctors and people from all over the tech and biotech and energy industry in Silicon Valley took it too.

What I learned for my $500 was how to pitch an idea better. I learned how to present ideas for grant proposals and funders in ways that, while not making them bulletproof, significantly increased their chances of funding. Originally I resented business types—the way they dressed, the words they used, and the Bluetooth earpieces they wore at the airport. I thought that they exploited creatives—the scientists—in technical organizations. The interplay between business and creativity was lost on me before I took this course.

The first reading we were assigned was a piece about the Motorola Iridium phone, a $5 billion failure. In the mid-1990s, this phone had been advertised as the next big thing. You could take it anywhere. It was a satellite phone, and Motorola had launched satellites for its specific use. The phone? Insanely expensive. Still, when it was conceived, it was set to connect Americans in a way that they had never before been connected.

Except that terrestrial cell communications began proliferating at the *exact same time*, rendering this expensive, overwrought project useless for the vast majority of potential users. Anyone who ever needed to make a mobile phone call could now do it on a terrestrial phone. So why invest in something so much more unwieldy and expensive? The case study of the Iridium phone is taught to MBA students as an example of what happens when the scientists and engineers get too much power within a company. As a postdoc who never had to administer an organization, I bristled at the suggestion. However, I came to realize that bringing a product to market requires not only technical sophistication but also judgment, an awareness of timing, cost, and societal need. This conclusion—perhaps obvious to individuals who don't spend all day, every day working with futuristic ideas in a lab—is still hard to swallow for me. However, it is probably the reason I have been so selective regarding the projects in which I have decided to pursue patents. Fingers crossed, but none of the startups in which I have been involved as an advisor, inventor, or investor have yet failed.

My team-based project, which culminated in a pitch to a panel of real-life venture capitalists, originated with one of my team members, Dhruv Kazi, MD, now professor and head of cardiac critical care at Harvard's Beth Israel Deaconess Medical Center. Kazi's idea arose from his experience in the ICU as a cardiology fellow. When a central venous catheter is inserted into one of the large veins in the neck to deliver drugs or fluids, the guidewire, which resembles a thick metal guitar string, sometimes breaks off or gets lost inside the vasculature. I proposed and designed a guidewire and catheter system that was impossible to lose. While Kazi, our other team members, and I went our separate ways at the end of this course, the loss of the guidewire still haunts ICU trainees.

My time in the program left me with questions about my future career. I remained interested in the academic route but also wanted to explore opportunities in business and industry. I was still a postdoc, but my fellowship from the intelligence community program was ending soon, and if I wanted to stay for a third year, Bao would have had to pay me from one of her grants. (A postdoctoral position doesn't have a defined length. You stay until you get a job, quit, or get fired, or the money runs out.) I applied for several positions, including a summer camp hosted by the Boston Consulting Group for aspiring management consultants. Dina and I were both rejected. I submitted applications to

Apple, IBM, and a startup doing work similar to my PhD research. None responded to me.

Meanwhile, I started to look my CV over once more. *This looks like a successful faculty application,* I said to myself. In the summer of 2011, I began working on research proposals, two pages for each of my three best ideas. I wrote my teaching statements. Back then, everyone was asked to write diversity statements, so I wrote those too. I naturally needed three reference letters, which I obtained from Jim Panek, George Whitesides, and Zhenan Bao.

The part that faculty applicants spend most of their time on is the research proposal. Notwithstanding, it's probably the part that the committee spends the least amount of time reading, because it takes a long time to read any dense text. Furthermore, the reviewer knows the short-listed candidates are going to present the research proposal in a more easily digestible slideshow at the on-site interview. What the reviewers are looking for to come up with the shortlist is the number of publications and the reference letters, particularly those from the PhD and postdoctoral advisors.

I applied for programs in chemistry, materials science, and chemical engineering across twenty-eight universities. Each position attracts two hundred to three hundred applicants and generally short-lists five candidates to interview. I was invited for seven in-person interviews: chemistry at UC Irvine and chemical engineering at the University of Maryland, CU Boulder, UT Austin, and the University of Washington. I had a second interview at UT Austin, held simultaneously with my interview in the chemical engineering department, during a grueling three-day process. This second interview was for a professorship in the Texas Materials Institute, or TMI, which was apparently not named with an eye toward acronyms that could be used for fun.

At the time, my dream position was at the University of Texas at Austin. One of my mentors in graduate school, Michael Dickey (now a professor at North Carolina State University), had done his PhD there and loved it. The city of Austin is the self-proclaimed "Live Music

Capital of the World," and the university had a strong community in organic materials science. The chemical engineering department faculty directory was a who's who of every subfield of chemical engineering—it seemed to have more members of the National Academy of Engineering per capita than any other department. Most importantly, although Dina grew up overseas in Saudi Arabia, she was born in Austin, she went to high school there and UT Austin for college, and her family and many of her friends lived there. If we ever had kids, we reasoned, we would have a built-in community. I had been short-listed for both positions, the one in chemical engineering and the one in materials science, so I liked my chances.

I didn't, however, like how I comported myself during my interview. It was at times a recapitulation of the world-class choke I had experienced at seventeen when I auditioned for music conservatories on trombone. I had twenty-three one-on-one meetings, two research seminars, and half a dozen meals over the course of three days. At my seminar I answered questions defensively and without much conviction—a characteristic for which Bao had criticized me during my postdoc in her lab—and I fell flat in my interview with the chair of the chemical engineering department, Roger Bonnecaze, in whose presence I felt like an imposter. Bonnecaze was the future dean of engineering at Texas, and I was not an engineer of any kind, but a chemist.

I suspect that the defensiveness in answering questions arose from insecurity and immaturity and that the lack of conviction arose from both fatigue and stress. I felt like I had been more successful at Maryland, Colorado, Washington, and Irvine, but when they all sent me rejections, I knew that I would also be rejected from Texas. So desperate was I that I sent a fawning email blast to the faculty at Texas with whom I felt I had made a connection. I immediately got a phone call from Nicholas Peppas, a superstar senior faculty member and founder of the program in biomedical engineering, who encouraged me, essentially, to take a chill pill.

The rejection from UT Austin came from Bonnecaze in a single email. In a final act of desperation, I asked him if this rejection was for both positions for which I had interviewed, chemical engineering and materials science. He assured me that it was. When I returned home to Palo Alto, I learned that a friend of mine in the Bao group, Guihua Yu, had received the offer from Texas and intended to take it.

In the end I wasn't too surprised. I was a fake chemical engineer. I had no coursework in the discipline, and my research portfolio was a disorganized mess: antifungal compounds, jelly roll solar cells, and stretchable conductors. This disorganization was not something Bao had brought up as a red flag in my portfolio, possibly because it was too late to do anything about it. At any rate, I had been rejected from six programs in a row. I remember sitting at my desk in our Palo Alto apartment, listening to "I Dreamed a Dream" from *Les Misérables* and sobbing melodramatically at my desk, like a heartbroken teenager.

I have now sat on both sides of the faculty interview. I realize now that part of the job of the interviewer is to make the candidate feel wanted. After all, a strong candidate may have multiple offers, and flattery might tip them your way, should you make them an offer. If you don't, well, it didn't cost you anything to be nice. For the candidate, getting a rejection after being buttered up so indulgently is especially gutting, like being dropped suddenly from the Tower of Terror. In one memorable conversation, I was told by a prospective colleague, "I'm not a superstar, but *you're* a superstar." One department chair asked me what he had to do to convince me to go there instead of UT Austin, as if I had offers from both places. In a couple weeks I would receive rejections from both. At any rate, it was all but over.

I had one interview left. And this position was not in chemical engineering but in the Department of NanoEngineering at the University of California, San Diego. What even is nanoengineering? Founded in 2007 in the midst of a boom in all things "nano"— nanotechnology, nanomedicine, and nano-enabled home beauty products—the Department of NanoEngineering hired all comers. Its faculty featured an amalgamation of chemists, chemical engineers, and materials scientists. It was founded on the philosophy that the frontiers of discovery in all fields of technology converged to the scale of nanometers. Computer chips, medical advances, batteries, solar panels, aircraft, and building materials all required the ability to engineer materials at the nanoscale. This department was going to train a new generation of interdisciplinary workers to bring this future closer.

So, I flew an hour south to San Diego. On the drive to campus, I was enthralled by the palm trees, the sunny skies, the canyons, and the ocean view. Even better than the Bay Area. When I arrived on campus, it was clear that the department had become proficient and maybe even a little jaded by interviewing candidates. The department had been founded a few years earlier, and they had worked hard to build a strong core of faculty quickly. They put me right to work. I gave a single job talk combining elements of my past work and what I proposed to do in the future.

In my lecture, on which all depended, I pulled out all the stops. My approach was like a performer on stage at TED: I walked around. I looked at people in the audience. I used—as Whitesides had suggested—"verbal punctuation" and played up the "nano" in my PhD research. I met with many impressive faculty members: Joseph Wang, who made nanorobots that could enter the bloodstream, reminiscent of those in *Fantastic Voyage* or *The Magic School Bus*; Shirley Meng, whose every paper was bringing high-performance lithium-ion batteries closer to reality; and Liangfang Zhang, who used the coatings of red blood cells on therapeutic nanoparticles so they could evade the patient's immune system—a cloaking device, but on the nanoscale.

I told the founding chair, Kenneth Vecchio, who would have the biggest say in hiring, that I intended to "kick ass." I would bring in millions of grant dollars and publish hundreds of papers. These were statements of insecurity and desperation that make me shudder with embarrassment when I recall them now. After an agonizing three-month wait—when I had no other options, as my postdoc fellowship would soon expire—I received another email from Vecchio. This one I received while working in the Bao lab, wearing gloves, goggles, and a lab coat, transforming glass slides into solar cells. In short, Vecchio told me that I got the position at UC San Diego.

While the feeling I had upon passing my PhD defense was one of relief, this was one of validation and pure elation. I would learn later that among members of the committee, I was the first choice of only Vecchio, though I was the second-place choice of every other member, who gave their first-place votes to candidates closer to their research areas.

Off I went to San Diego.

Chapter 8

Dina could not move with me from Palo Alto to San Diego, as she had to stay behind for three months to finish her postdoc, a situation that's not uncommon for academic couples. So, it was just me and Nunu. From my perspective, if there was ever an ideal time to be alone, it was the first few months of an assistant professorship, when I was expected to get my lab and teaching off the ground.

When I started at UC San Diego in 2012, the job came with a startup package of $640,000. These funds were to be used for equipment to outfit the lab, my own salary during the summer, and salary support for PhD students and postdocs. UC San Diego also offered the bonus of a below-market-rate faculty housing complex, provided for faculty and staff. The complex had an outdoor pool, and it was about two miles from campus. I didn't have a car when I first started, because Dina had kept it in Palo Alto, so I would walk to campus, accruing four miles in steps on my round trip each day. It was an easy and lovely way to get exercise, just another thing to love about California. Sure, Boston had been great, those runs along the Charles, but here was California's ever-sunny weather, hardly a drop of rain in sight, swaying palms, beachy vibes, fit folks in every direction, and a happy-go-lucky kind of atmosphere that made you understand, deep in your bones, why people hated the weather back East so much.

Living off graduate and postdoc salaries in some of the most expensive cities in the country meant we had no savings at this point in our lives. We especially appreciated that UC San Diego offered an incentive whereby faculty members could spend part of their research startup money on the down payment for a home. For us, that was nearly $40,000 after taxes. At the time, the San Diego housing market had not yet peaked, though it was still expensive. In 2013 we bought a house for $645,000, four miles from campus and within a residential neighborhood. Using UCSD's competitive mortgage program, we found ourselves in the right place at the right time to acquire equity through real estate. The program offered a forty-year term, a 10 percent down payment, and an adjustable rate that began at 3 percent and was not allowed to rise more than one percentage point per year. We had no idea how sweet this deal was. Our monthly payment? A steal for San Diego: $2,200 a month, for a 2,100-square-foot home with a pool, ten minutes from campus.

When I started at UC San Diego, I was just a new faculty member blindly stumbling through a nascent department and trying to create my own unique research group. Famously, professors at research universities are hired for their skills at the lab bench or computer terminal. They are not trained to teach, manage budgets, negotiate with vendors, hire personnel, or establish a climate of trust and openness in the laboratory, and there is no playbook to follow. Those who achieve success in this world learn how to do these things on the job.

My assigned laboratory space was in the new Structural and Materials Engineering Building, which had opened a month before I arrived. My lab was about eight hundred square feet. It sported two chemical fume hoods, two sinks, high ceilings with exposed utilities, and wall-to-wall benches with a large center island. Its most striking feature was a view overlooking the picturesque Pepper Canyon, a former wildlife space that was later infilled with concrete to accommodate a performance venue and a high-end dormitory.

I joined UC San Diego during a period of high growth. Chancellor Pradeep Khosla raised the campus's profile meteorically during my twelve years there. A student body of just over thirty thousand when I started increased to forty-six thousand at the time of this writing. During my time, UCSD was the second-most-applied-to university in America. The first was UCLA, and the top five are all in the UC system.

The Department of NanoEngineering was singular within the United States. A handful of other departments featured the word *nano* in their names, but their programs were oriented toward technical training:

educating workers who could execute in the semiconductor industry. None of these programs existed at a place with the scholarly heft of UC San Diego behind it. My campus may not have been in the top five of *US News and World Report*, but it was top five in another, possibly more important metric: the annual research budget, which was nearly $1.5 billion at the time and is much more now.

When the nascent Department of NanoEngineering was hiring faculty members, they interviewed PhD holders with all kinds of academic training. The feature common to all these researchers was an interest in materials whose function arose from phenomena occurring at the nanometer scale. Around half the personnel hired had research interests in biomedicine: drug delivery for cancer therapeutics, devices for high-resolution medical imaging, and, like me, the development of materials that could be used in new types of devices for physical therapy and behavioral interventions. There was also a strong cohort of faculty members working in areas related to energy and sustainability: batteries, solar cells, and catalysts. When I joined the department, there were twelve faculty members; now there are thirty.

When an assistant professor starts their research program, they are doing so from scratch. In addition to having a completely empty lab, I had no personnel. The university keeps track of the revenue generated per square foot of lab space. The per capita research expenditure for a faculty member in engineering was $800,000 per year. While my department chair was never so explicit, I knew that I had to bring in $1,000 per square foot of my laboratory each year just to make the math work—to justify my presence on the faculty. Of course, the job of a professor is to do *scholarship*. However, what constitutes scholarly activity varies from field to field. It is also difficult to evaluate, even between professors in the same department. Research expenditure, at least for STEM faculty, is a serviceable proxy for intellectual merit. It is easier to measure than academic achievement, as funds flow out of the lab through a single pipe. While a focus on funding may be reductionist, it has the advantage of leveraging the expertise of outside faculty members, as they are the ones reviewing grant proposals and scientific publications, the documents that undergird successful proposals for funding.

To recruit for my new lab, I sent out emails to senior professors encouraging them to ask their PhD graduates to consider my lab for their postdoctoral work. As is my modus operandi, I spent hours reviewing each email—the first communication they would receive from me as a professional colleague. For PhD students, I relied on the department to advertise my new lab.

The first student I hired was Adam Printz, who had just completed his MS in nanoengineering and wanted to continue to the PhD. He found out through the grapevine that a new assistant professor—me— was starting in the department. These days, Printz is a tenured associate professor of chemical and environmental engineering, and materials science and engineering, at the University of Arizona. Printz was a year older than I was. An undergraduate major in finance, he spent his early years running his own wealth management and investment service. He also completed all the required coursework for a chemical engineering degree. Despite his modest seniority to me, he was nervous about our first meeting, and told me so. I looked at Printz and thought, *A kindred spirit!* Other students had applied to the lab, but I hired Printz because of his maturity. Having run his own business was a huge plus.

My next hire was Don Garrett, whom I brought on as a postdoc. Garrett came to me based on the recommendation of Professor Craig Hawker, a famous Australian polymer chemist who made his name at IBM, inventing several of the materials used in the manufacturing of semiconductor chips that we now take for granted. Hawker had become a professor at UC Santa Barbara, where he was Garrett's PhD advisor.

Garrett was a bit of a firebrand, and he made a lot of enemies (particularly within the equipment services team) during his short time with me; he remained in my lab for only one year. Garrett had a real distaste for one guy specifically, Rich. Rich never made appointments. He showed up on an unpredictable schedule and bragged about the size of his utility knife. Garrett couldn't stand it. I found out, after a while, that Garrett had been fighting with Rich—via email, as is the scientist's way—raising issues about equipment and about showing up on time. That was when Ken Vecchio got involved.

By the time Rich's complaints elevated to Vecchio, Garrett had one foot out the door. I assured Vecchio that it would soon not be a problem and apologized for letting things get that far. Garrett soon left to make battery materials for a series of well-known companies in the electric vehicle space. While his tenure in my group was short, he helped seed a culture of directness in the lab that persists to this day.

Tom Connelly was the next to join us. He was, to put it bluntly, a character. A fan of electronic music and a regular attendee of the Burning Man festival, Connelly had been kicked out of his first undergraduate institution, in conservative Ohio, for possessing marijuana. He was a physics major and contributed to the overall culture of my lab, which followed many of the tropes of the California art scene. My team of researchers embodied a work-hard, play-hard ethic, which was well known by the other students in the department, though its excesses were not known by the faculty and administration. Garrett had left behind a high-powered stereo system in the lab, which was always blasting. You could hear it down the hall, and I encouraged this free-spirited culture. The contrast offered by my lab—compared to the straitlaced and serious identity embodied by most academic engineers—attracted a lot of applicants, and it was always hard to turn down eager students because, like all academic research groups, my space and funding were limited.

One prospective PhD student, two years my senior, was a political exile from a former Soviet republic. As a student in his home country, he had been part of a pro-democracy group, a perilous thing to belong to. Weeks before graduating, the government threatened him and his family. He applied for political asylum in the United States, working odd jobs: demolition, fast food. After achieving another four-year degree in biomedical engineering from a large public university on the East Coast, he joined the master's program at UC San Diego and found his way into my lab.

Back then, I taught a course that I used as a recruiting tool. Intermolecular and Surface Forces was designed for first-year graduate students and was a course I had inherited from a departing member of the department who had received a better offer from the University of Colorado at Boulder. My class had thirty students, three of whom ended up joining my lab.

I was nervous about teaching. Most professors complete only a couple dozen hours of leading a recitation section as PhD students before they are asked to teach a class of paying customers. Prior to becoming a professional educator, I had only limited experience as a TA at Boston University and Harvard. More importantly, the first class I was assigned to teach was a graduate class that I myself had never taken as a student. The course had used a classic textbook, *Intermolecular and Surface Forces*, written by Jacob Israelachvili, an Israeli physicist and chemical engineer from UC Santa Barbara, which defined the field. There was a good chance that several of my thirty students had used the book in a previous class. If I said something wrong, someone would probably know. I was so nervous that I would go into the lecture hall the night before each class and write out my entire lecture on the chalkboard. I practiced what I would say—jokes included—to an empty room so that I would feel prepared for the next day. I had heard about other faculty members preparing so meticulously. I had thought, at the time, that such preparation tactics sounded insane. Nevertheless, there I was, doing the exact same thing.

The content of the course was hard, and my knowledge bottomed out—a lot. In many instances, literally everything I knew about the topic was written on the board. I had no further depth of knowledge from which to draw. If anyone asked questions during those times that my keel was scraping the ocean floor, well, I wouldn't know the answers.

Finally, I hired Suchol Savagatrup, now an associate professor of chemical and environmental engineering at the University of Arizona. He's in the same department as Adam Printz. Both are now tenured and have won several prestigious federal grants.

Our lab was diverse. Academic labs in the US have historically included many students from China, India, and Middle Eastern countries—notably Iran, though less so under Trump. My group, though, was really unusual. Savagatrup was a dual citizen of the United States and Thailand. My political exile student was a dual citizen of the United States and a former Soviet republic. Throughout the years, we had several students who were dual citizens of the United States and Mexico. Although there is a high percentage of undergraduates of Mexican heritage at UC San Diego owing to the proximity to Tijuana, this group is severely underrepresented at the graduate level. In my group, it was properly represented. Julian, Daniel, Guillermo, Laura, and Brandon all earned their PhD in my group and are enjoying successful careers.

My criteria for accepting students into my lab were unusual. I had to know students for a long time and rarely just plucked one from the pool of fresh recruits to the department. Two-thirds of my PhD students either worked with me as undergraduates or began their program in one of my colleagues' labs and found a better fit in mine. Most came from nontraditional backgrounds. Some spent significant time working before going back to school for their PhDs; older students are quite uncommon at top graduate programs.

International students, dual citizens, and multilingual group members—who could have gone to other countries—studied in the US because our research universities are the envy of the world. Not only do we produce a lot of research, in terms of papers, but what we produce is innovative and impactful. More Nobel Prizes have been won by Americans than by citizens of any other country. Americans account for nearly half of the total. The vast majority of all scientists who have left their country of birth and gone on to win the Nobel Prize won it while working at US universities.

When I started my position, I was twenty-nine. Since many of my students had previous life experience before going back to school, we tended to be close in age. It wasn't until I had been in my job for ten years that I was finally the oldest member of my group. The camaraderie we enjoyed was unique and will never be replicated, as the average age of my students remains constant while I get older. In those first few years, we celebrated with beer and pizza every time a paper was submitted, every time a paper was accepted, every time a paper was published in final form. If one of our graduate students passed an exam, we all went out. It was not just a deliberate strategy to increase the group's cohesion and to set a culture of support in the lab; it was also fun for me personally. We had parties at my house. My pool became ground zero for social gatherings throughout the year. For a few years, there was an annual tradition involving my students coercing me into shotgunning an especially insidious liquid known as Bud Light Lime, a memory that still horrifies my wife.

Other instrumental people began to join our group. Sam Root showed up at a graduate recruitment event, evidently a bit of a savant. In high school and college, he concurrently read thermodynamics textbooks (for fun) and fantasy books (also for fun). After graduating from my group in a record three and a half years, he followed in my footsteps as a postdoc with George Whitesides and then with Zhenan Bao. I was of course proud of him for obtaining these positions. I was also gratified that my former mentors trusted my abilities as a mentor enough to hire one of my own students. With Harvard and Stanford now on his résumé, he recently started his own lab as a professor of polymer engineering at Case Western Reserve University.

Then came Brandon, a Mexican American student hailing from a poor neighborhood in East Los Angeles. Brandon was born the same year as me and also loved *Star Trek*. He was also a competitive, sponsored skateboarder. He did well in school and had his choice of colleges.

During his first week as an undergraduate at the University of Southern California, Brandon met his roommate, who, it turned out, was a heroin and meth addict. Brandon got into drugs, and not the kind that hippies and Silicon Valley types use to augment their creativity. He was kicked out of school. Over the course of five years, he served a total of one year in county jail as a consequence of five felony convictions. His mother all but disowned him, and he spent a handful of the intervening years living on the street, homeless. The judges could see something in this brilliant kid, but he was an alcoholic and an addict. He had one final conviction, and the judge gave him the option of two possible sentences: "You can either go to inpatient rehab and stay clean for one year, or you can go to state prison for five years. If you fail to complete rehab, you'll go to prison for ten years."

His mother sat in the gallery, mouthing, *Prison, prison, prison,* knowing he couldn't last a full year in rehab.

"I'll take the rehab," he said.

That was in 2007. After a year in rehab, Brandon had come to know all the other patients, 120 in total, most of whom have since relapsed, become incarcerated, or died. In contrast, Brandon became a PhD student in 2012, successfully earned his PhD in 2017, and is still clean and sober. He has a family. As a senior engineer at Intel, he contributed to over one hundred patented inventions.

My team was full of superstars and the envy of my colleagues. It didn't take long for us to start drawing attention to our work. In the four-year period from 2012 to 2016, my group was awarded funding from the National Science Foundation (NSF), the US Air Force, the Department of Energy, and the National Institutes of Health (NIH). We were discovering and inventing new types of semiconductor materials that had the properties of rubber. Potential applications of our discoveries were myriad. They included solar cells, batteries, and biomedical devices. Our papers were being accepted to highly visible journals, and I was being invited to give talks at prestigious international conferences. Two of my students were about to graduate and obtain competitive postdoctoral positions at Stanford and MIT, a third was starting his own company, and a fourth was Brandon, who was headed to Intel. Given the fast pace of research and the attention it was garnering, I applied for tenure.

The concept of academic rank within the university hierarchy seems designed to be misunderstood. At research universities, most faculty members hired are on the tenure track. A smaller number are hired primarily to teach, and rarely do these instructional-track professors have the opportunity to get tenure. A new faculty member on the tenure track is called an assistant professor. Despite the title, an assistant professor—sometimes referred to in jest as an "ass prof"—does not actually assist anyone, at least in the American system. They run a research group and teach, the same as any professor of higher rank. If an assistant professor fails to get tenure, they get fired. There is no middle ground.

Tenure is a form of job security designed to protect academic freedom. It means that your institution trusts you enough to explore controversial topics—as a bulwark against the ephemeral cultural zeitgeist. It also means that barring some financial calamity or significant run-in with the law, you will have a job for life and a prosperous retirement. Once tenured, you are promoted to associate professor. At a small number of universities, usually the more hoity-toity ones like Harvard, the promotion to associate professor doesn't come with tenure. For that, you must achieve the rank of full professor, or just "professor."

Professors of any rank can do research work in any area for which they can get funding and that vaguely resembles the mandate of the department (i.e., doing chemistry in a department of chemistry). However, when it comes to teaching, they serve at the behest of the department head (or chair). Most research-active faculty members in the US teach one course per term. The department head also controls laboratory space, the department's non-research budget, and service roles. Service includes what a parent might call chores—nonacademic work like recruiting students and faculty, determining grading policy, event planning, advising student clubs, and evaluating cases for tenure and promotion.

A traditional tenure-track position sees an assistant professor applying in the fall semester of their sixth year. The process takes an agonizingly long time, and the decision to grant tenure usually isn't made until late in the spring. An application for tenure—a tenure package—consists of a cover letter or summary statement, a research statement, a list of grants funded, courses taught, and teaching evaluations. The department solicits external letters from members of the scientific community. Your colleagues meet, write a recommendation, and take a vote, which goes into your file. Your file then makes its way to the dean's office, where the associate dean for faculty makes a recommendation to the dean. The dean—or usually an assistant—writes a recommendation that goes on to a campus-wide committee, which compares your tenure package to tenure packages submitted by others at the institution who have undergone similar milestones at similar career stages. Ultimately, your package makes its way to the provost, the chief academic officer of the university, who will make the decision. If your package gets through all the hoops, you get tenure, effective the upcoming academic year, which starts July 1.

My progress put me on an accelerated track to tenure, meaning that I was able to apply during my fourth year and began as a tenured associate professor during my fifth year at UC San Diego. I had several papers and around $3 million in grant funding, and I had already graduated three PhD students. This was considered a rare achievement at UC San Diego, and I knew it. I was still quite young, but now I had the security of permanence in my role. I could stay for as long as I wanted because I had job security and, with that, the respect of the word *tenure* attached to my name. Removing a tenured professor is a complicated, bureaucratic mess, which is to say, after my fourth year, I was in a safe position, or so I thought.

I obtained tenure in 2016, the year Donald Trump defeated Hillary Clinton to become president. The first Trump administration instructed the Justice Department to investigate professors with ties to China. Several high-profile scientists, including Charles Lieber, Harvard professor and chair of the Department of Chemistry and Chemical Biology, from which I obtained my PhD, along with Gang Chen of MIT, became the subject of investigations.

It seemed that the government was pursuing individuals with even incidental connections to China—traveling there for a routine scientific conference, for example. In 2016 I received a recognition by the World Economic Forum as an up-and-coming scientist and was invited to Tianjin for a large expo of talks by world luminaries. Summer Davos, they called it, as the sister event of the famous expo each winter in Switzerland. Later I traveled to Shanghai to give a talk at Fudan University. Mind you, such invitations are not unusual. I traveled internationally several times per year as a professor. The goal was indeed to exchange information with the international community. Despite my funding from the Air Force, nothing I was doing was classified, and all my results were destined for publication, for the widest possible dissemination.

One afternoon I was meeting with my PhD student Julian. There was a knock at the door. I always ignore knocks when I'm in a meeting. I have an open-door policy otherwise. When the knocking persisted, Julian opened the door. A man and a woman, dressed to impress, were at the door.

"Professor Lipomi, may we speak with you?" one asked.

"I'm in a meeting with a student. Can you come back later?" I replied. I figured they were trying to sell me pipette tips.

The man opened his wallet and showed me his FBI credentials. "May we speak with you *now?*" he insisted.

Julian, jaw dropping as if to mouth, *Oh, shit!,* said that he'd better go. The agents let themselves in and shut the door behind them.

Despite my fears that I had perpetrated a crime worthy of federal prison, the conversation was mundane. The agents gave the impression that they were interviewing people at random, trying to create leads in support of the Trump Justice Department's China Initiative. The agents asked me where I had been in China and for what purpose, whether I ever left my electronic devices unattended, and if I knew anyone involved in "foreign talent programs." They had nothing, as I had done nothing, and I told them so. I assured them that my America First credentials were unimpeachable. In the end, the male agent gave me his business card and left my office. Evidently, several professors across campus were similarly accosted by the FBI. We were later told not to talk to these agents without first contacting campus lawyers and waiting for them to arrive. A few years after President Biden took office, the China Initiative was terminated.

A researcher with a public profile quickly realizes that science has its critics on both the Right and the Left. The same period as the height of federal harassment of researchers during the China Initiative was the height of callout culture and the "Twitter mob," people who are primed to jump on you for perceived slights. Well, I stepped in it, big time. My lab had started working on new materials for haptic devices that stimulate the sense of touch. To make a long story short, we made a glove that could control a virtual hand using a cheap type of conductive paint. When the paint dried, we found that it was a better motion sensor than the much more expensive sensors other labs were using. We wanted to get a quick paper out, in the open-access journal *PLoS One*, so that the community could start benefiting from our finding. The paper was led by Tom Connelly, my EDM enthusiast. Connelly and I decided to use the glove to render the letters of the American manual alphabet. This signed alphabet is used by Deaf people to render loanwords from English into American Sign Language (ASL), which is called "fingerspelling."

Ignorant to the fact that some Deaf people and linguists would rightly bristle at the characterization, we called our rendering of gestures into letters on a screen a "translation." We also gave the paper a cringeworthy title: "The Language of Glove: Wireless Gesture Decoder with Low-Power and Stretchable Hybrid Electronics." UC San Diego published a press release on the paper, and several media outlets, including local TV, filed their own reports. With an eye toward a story graspable by the public, they spun our project as an accessibility device— a device used to make it easier for hearing people to understand deaf people.

After several of these stories were reposted on Twitter, I received a message from my dean. "It seems your glove paper touched a nerve," he said, attaching a scathing letter he received from the president of a prominent American society of professional linguists. This letter accused my team of cultural appropriation and ableism. Our alleged missteps included conflation of fingerspelling with the whole of ASL, not including linguists on our study, and suggesting that Deaf people should bear the burden of wearing our clunky device. The letter gave us an ultimatum: publish a correction or other acknowledgment of our missteps, or the group would go public with their accusations.

I was enraged at being called ableist and incensed on behalf of Connelly, who was just a student, had recently graduated, and was trying to get a job. I was also mad at myself for having such a blind spot. I published an addendum to the article, acknowledging our lack of cultural awareness and providing corrections to the text.

I thought I had paid my penance, but months later I received an email from a journalist doing a story for *The Atlantic*. The story he proposed was about my paper, how sign language gloves don't help Deaf people, and basically how engineers are cultural troglodytes. I found that he had earlier tweeted my paper with the comment, "Assign this story to me, how engineers keep fucking this up." Nevertheless, I agreed to cooperate with him. I was quite sure that those criticizing me on Twitter did not read the paper and were instead reacting to popular articles about it. For the story, I agreed only to be interviewed on background, under the condition that he cite my own published addendum to the paper. When his article came out, it characterized engineers as uncultured buffoons who never set foot outside the lab. The article appeared on the landing page of *The Atlantic*, but I didn't hear anything about it afterward. Someday, I hope to be covered in a publication with a similar stature for more flattering reasons.

While I may have had my cultural blind spots, I was acutely aware that all my PhD students in the first few years were men. It didn't help that two women graduate students were briefly in my lab and departed under unfortunate circumstances unrelated to their gender. One ended up resigning from the program. Another was in the lab for about six weeks and then quit. When eighteen hours later she told me she had made a mistake and asked to rejoin the lab, I said no. I didn't see evidence of commitment to a five-year program. I had let go of male students for similar reasons. Despite a reputation I feared we were developing of unfriendliness to women, some took a chance.

First, there was B, who was Turkish but had obtained her bachelor of science in the US, at Johns Hopkins. Then came Laure Kayser, a postdoc of French origin, now a tenured professor in the Department of Materials Science and Engineering and the Department of Chemistry and Biochemistry at the University of Delaware. Kayser stayed with us for three years. Rachel, a brilliant postdoctoral chemist who wrote a thick phonebook of papers while raising three kids, was with us for six. And Laura, a phenomenal electrical engineering student who spent half a year in Ecuador teaching English, and Allison, who joined during COVID and is now a consultant. Our culture became irreverent and funny. We were like a giant family, particularly in our most productive years, from 2016 to 2019, right before the pandemic eclipsed everything.

My lab evolved into a community that mixed technology and art. It turned into a space for things that were bigger than just the basic science of an arcane topic with which we began—the mechanical properties of semiconducting polymers. My research group was already unusual in terms of its breadth of research. Perception research. Psychology. Making devices that make users literally feel virtual shapes and textures. And a lab can be a space to explore how research and personalities combine. I learned in those tender first years as a faculty member what it means to cultivate a lab culture.

Chapter 9

Historically, among the most damning epithets one scientist could cast upon another is that of "popularizer." Scientists, the thinking goes, should not cheapen real discovery by turning it into a parlor trick or wrapping it in fantastical language. (And they probably shouldn't have YouTube channels!) There may be a fine line between excitement and hype, but going overboard in your enthusiasm for a discovery or invention can have real consequences. For example, funding agencies—and even other researchers—may come to believe that the endpoint in a particular field has already been reached. The bar for getting careful, sometimes necessarily incremental, work then becomes impossibly high. "Hasn't this already been done?" is the distillation of a common critique in the peer reviews of grant applications. Credit sometimes goes to the first to claim rather than the first to execute.

Among science communicators who are active scientists, I like to know a little bit about their perspective—how their professional research informs their worldview or allows them to speak with authority on topics outside of their professional niche. So, I wanted to give you a little peek into my research. Before I do so, I want to emphasize that essentially all the work I'm about to describe—and much more, which is covered in the 140 peer-reviewed publications I have authored—was executed in the laboratory primarily by my students and postdoctoral scholars. My role has been to provide the seed of an idea, maintain scientific rigor, offer mentorship, and secure funding for the work.

As far as lab work goes, the last time I spent any significant time running reactions or analyzing samples was during my postdoc at Stanford, which ended in 2012. At UC San Diego, and now that I'm at the University of Rochester, days can go by without my having set foot in the lab. It's not just laziness or busyness on my part, but also because poking around my students and postdocs while they work stresses them out. I'm working on ways of being present without being invasive. The fact is, I miss the sounds, the smells, and the way time flies—truly engaged—optimizing the conditions of an experiment.

I often wonder what life would be like if I returned to the lab. Nothing is really stopping me, except possibly the frustration of a failed experiment or the tedium of repeating experiments to ensure reproducibility. It is also true that my skills are now more suited to higher-level tasks, such as research strategy, mentoring, and fundraising. I *do* miss the manual aspects of lab work: the way time flies when you're shooting the shit with your labmates, and accidentally working through lunch- and dinnertime because you're too engrossed in your experiments. I currently try to scratch the itch of doing something manual through cooking, home repair, and maintaining my collection of vintage trombones.

The students in my lab have run the gamut in background and training: chemists, materials scientists, and all branches of engineering and computer science. What gives me any edge is that I am eager to work with scholars far afield from my home base in organic chemistry. These fields have included psychology, neuroscience, music, and visual arts. For example, I have had many productive collaborative projects with V. S. Ramachandran, a trained neurologist and famous cognitive scientist who previously appeared on *Time* magazine's list of the one hundred most influential people.

Ramachandran is best known for his invention of mirror therapy for the treatment of phantom limb pain. A common treatment at veterans' hospitals, a mirror box is used to manipulate circuitry in the brain to reduce pain that appears to emanate from an amputated limb. With my group's expertise in materials science, along with Ramachandran's expertise in psychology, we discovered many of the parameters that make a material feel smooth, soft, and wet. We also developed materials that make virtual worlds feel more realistic for future applications in medical training and remote operations in dangerous environments such as space, the deep ocean, and wildfires. Working with Ramachandran and his brilliant graduate student, Dr. Nicholas Root, was a true honor. Our experiments showed that you didn't always need fancy equipment to do science. Ramachandran called this kind of low-resource work "Victorian."

On the opposite end of the spectrum of sophistication, I have had a productive collaboration with Ardem Patapoutian, who in 2021 won the Nobel Prize for his discovery of the cellular structures—called Piezo proteins—that enable the sense of touch. Granted, my group's contributions to this work were minor in the grand scheme of things, but supportive roles can be crucial in providing a key insight that propels a discovery or invention. The difficulty of making discoveries that are completely new and transformative increasingly requires contributions from a broad swath of scientific disciplines, which is why papers published in *Science*, *Nature*, and *Cell* commonly have twenty-five authors or more.

When I started my independent research career, I consciously chose a range of projects from incremental and "safe" to high-risk, high-reward. It was like assembling an investment portfolio: bonds for security, stocks for risk. Incrementalism has a bad rap in the popular press—people want "disruptive innovation" and "blue-flame thinking." But someone must do the steady work of finishing ideas, developing complete knowledge of a material, molecule, mechanism, or effect.

Incremental science creates discipline in trainees and long-term impact in society. A single one-off paper in *Science* or *Nature* is good for a scientist's career, but it rarely changes the world. Without years of rigorous experimentation, ideas are too risky for entrepreneurs, companies, and investors to take forward. For professional academics, it is also practical to maintain a steady "engine" of work that produces papers, brings in grant funding, and trains PhD students for the workforce. Whitesides used to tell us that a discovery must be developed until it becomes boring before it's ready for commercialization.

The core of my own early career—the work that became my engine and defined my reputation—was focused on the mechanical properties of organic semiconductors. These are materials that absorb visible light and conduct charge in a controllable way, just like traditional semiconductors. What makes them organic is nothing to do with farming practices but rather that their molecular backbones are made of carbon atoms, as opposed to inorganic compounds such as metals, minerals, ceramics, and salts.

The most common place you'll find these materials is in organic light-emitting diodes (OLEDs), which are used in all modern iPhones and in the best high-definition televisions. The global market for OLED technology is now approaching $100 billion. Other applications under development include solar energy, batteries, and biosensors.

Organic semiconductors come in two classes: polymers and small molecules. Polymers are long, chain-like molecules made of thousands of atoms, while small molecules are one hundred to one thousand times smaller—comparable in size to drugs such as acetaminophen or ibuprofen, or to simple dyes such as food coloring. To continue the analogy to dyes, many small-molecule semiconductors are vividly colored, one of the aesthetic rewards of working in the field. Current OLED displays are based entirely on small molecules, though polymers promise simpler processing—potentially by printing—and better physical durability, including flexibility and stretchability.

For many years the mechanical strength of these materials was taken for granted. Together with a small group of researchers worldwide, I showed that most were in fact fragile. Understanding how molecular structure affects strength and compliance became essential if these materials were ever to see broad practical use. Over the next decade my group produced a large body of work dissecting every relevant structural factor: chain length, atom placement, molecular packing, and thin-film morphology. Eventually we used our insights to make materials that were stretchable, conductive, and biodegradable.

Our long-term vision was a solar "tarp": a deployable panel for space missions, military operations, and disaster relief that could unfold to the size of a dance floor yet fit into a backpack. We did not fully achieve this ideal, but we did demonstrate several firsts—such as solar cells that could be laminated onto skin like a temporary tattoo, and the first fully rubber-based solar cells. Collectively, my work has been cited around twenty thousand times, becoming part of the foundation for further advances in the field. We have used the fundamental knowledge created by this work to make strides in other areas as well. In collaborative projects with neurologists, we are developing soft implants for the brain and scalp. These devices are capable of interfacing with the brain for the treatment and monitoring of neurological conditions, such as epilepsy and traumatic brain injury.

According to the National Center for Engineering Statistics, over half of the funding for research and development allocated by the federal government in 2023—approximately $60 billion—went to life and health sciences, nearly all of which is directed toward improving the lives and well-being of all Americans. That money was allocated through both grants and contract funding. Grants and contracts are slightly different. Grant funding is offered through large-scale federal departments such as the NIH, and the purpose is to fund research. Federal spending on health-related research, for instance, is permitted under the Public Health Service Act. Contract funding, where a specific good or service is paid for by the federal government, operates much as it would in the private sector. Universities help to fulfill national needs, and the government supports the research. NASA's agreements to test prototypes, for example, are structured as contracts.

In 2023 UC San Diego was the recipient of $1.08 billion in federal research and development funding. This amount places the university among the top five recipients of federal funding. This is a business in which the spoils are highly concentrated. The $60 billion total federal annual research budget is distributed to a few thousand colleges and universities, of which the vast majority goes to the 150 or so R1 institutions. A smaller portion of this money supports government-run labs such as the NIH facility in Bethesda, military research labs, and Department of Energy facilities.

Since starting my independent position in 2012, I have received about $10 million in government research grants. This total is divided roughly equally among the Department of Defense, the NIH, the NSF, and both the state and US Departments of Energy.

Grants are awarded through a competitive process. A principal investigator—usually a professor—conceives an idea, writes a proposal, and submits the proposal to the agency. These proposals are written in response to documents called funding opportunity announcements posted on the agencies' websites and updated every few years. Each proposal is reviewed by peers—other professors who receive funds from similar programs. These peer reviewers comment on criteria such as the significance of the problem, the innovation and approach of the solution, and the quality of the investigator and research environment. At the NSF, proposals must also include a section on "broader impacts," which for many years required investigators to explain how their work would contribute to broadening participation, especially by those belonging to groups underrepresented in STEM research. These groups include women, racial minorities, veterans, and low-income students of any race or ethnicity. This expectation has been eliminated under the second Trump administration.

The criteria on which proposals are judged vary by agency. Clinical trials must demonstrate rigorous statistics, human-subject protections, and authentication of biological samples. Defense Department proposals must tie into national security and warfighting readiness, often requiring investigators to stretch—or at least prognosticate—the relevance of basic science to America's defense aims.

Success often depends on relationships with program managers, who play an outsized role in shaping their portfolios. As a postdoc in Zhenan Bao's lab, I was introduced to Dr. Charles Lee, program manager of the Air Force's organic materials chemistry program. Lee's portfolio of funded researchers included a Nobel Prize winner, Alan Heeger of UC Santa Barbara, who was part of a team that developed the first electrically conductive plastic materials. Lee was a notoriously hard man to read, but I secured two grants from him before he retired.

The first project—the one that got me started—was written in my Palo Alto apartment while I was still a postdoc, just before beginning my faculty job at UC San Diego. The idea, which became the basis for my work on the mechanical properties of organic semiconductors, received positive reviews. You are not supposed to know who your reviewers are, but years later one of my mentors at UCLA, a senior professor of materials science, let slip that he had been one. The grant was valued at $360,000 over three years, enough to support about one PhD student, covering stipend, benefits, tuition, and materials.

There is also the matter of indirect costs, or overhead, which has been in the news since it became the bugbear of Elon Musk and DOGE early in 2025. Overhead covers expenses such as utilities, staff, and shared facilities. The government pays the university an additional percentage because it would be too inefficient to itemize every cost. At the time of this writing, standard overhead rates are 50 to 60 percent of direct costs for most universities. Indirect costs are not taken as a percentage of the total grant; rather, they are assessed as an extra amount (i.e., 50–60 percent) charged on only certain budget items. Tuition remission (the benefit granted to a student for not paying tuition but flowing to the school) and capital equipment purchases are excluded from the calculation. Once everything is accounted for, about 30 percent of the total value of a grant ends up going toward overhead.

There has long been debate over whether research earns or loses money for universities. The answer depends on how active a department is and how the university's budget model works. Specialized research institutes, such as Scripps Research in La Jolla, which has no undergraduate program, often have overhead rates closer to 70 percent. Universities with undergraduates can offset some of their overhead costs with tuition revenue, so they need to charge less for these indirect costs. Some departments, such as the highly ranked Materials Department at UC Santa Barbara, attract so much research funding that they do not need an undergraduate program to satisfy the financial demands of their campus.

Peer reviewers on grant panels are not expected to scrutinize budgets closely. Their job is to judge whether resources are reasonable, not to use the budget as a scoring criterion. Instead, panels focus on the science and the investigators' ability to carry it out. Reviewers submit preliminary scores individually, then meet as a group to further debate the proposals. In the process of scientific review used by the NIH, only the proposals receiving preliminary scores in the top 50 percent are discussed by the review panel. The rest are "triaged" and instead of getting a score get a rating of "not discussed." No two words terrify a researcher in the health sciences more.

Researchers argue about whether the peer review of grant proposals rewards true innovation or just incremental progress. The adage is that you get the grant for the work you already did, since large NIH grants, called R01s, lean heavily on preliminary data. Outsiders sometimes conclude that millions are being awarded for finished projects, but in practice the money supports new experiments that will fuel the next application.

The NIH also runs high-risk, high-reward programs. In 2014 I outlined a proposal that became my NIH Director's New Innovator Award. The proposal scored well, an 18 out of 90, with the lowest (best) score being a 10. This proposal was sent to the White House Office of Science and Technology Policy and was later recognized by the 2019 Trump White House in the form of the Presidential Early Career Award for Scientists and Engineers (PECASE). Between the initial grant and the supplement I received from the NIH after the recognition from the White House, the total grant was worth $2.8 million. The topic supported my work on biocompatible semiconductors for implantable sensors. Awards of this scale with so much built-in flexibility in the use of funds are rare. The more common mechanisms for exploratory work are smaller grants, called R21s, from the NIH or single-investigator grants from the NSF ranging from $300,000 to $450,000.

Professors frequently complain about the grant system. To submit a proposal, a professor takes their best idea and uses all their free time on evenings and weekends for a month to write it. Then, they face a success rate of 15 percent or lower. Rejections arrive months later, sometimes with perfunctory reviews, which we complain about. Yet when we sit on panels ourselves, many of us conclude that the process is reliable because the strongest proposals usually do rise to the top. Are many quality proposals left in the bucket, sometimes explicitly labeled as "fund if possible"? Absolutely, and I have been in that boat many times. During one depressing stretch early in my career, I had eleven consecutive submissions declined by the NSF. Later I had three in a row funded. Ask a professor any day of the week how their day is going, and their answer is going to be affected by the outcome of their most recent proposal.

Some criticize science for failing to achieve breakthroughs at the pace of prewar history, pointing especially to slowing productivity since the 1970s. It is true that singular discoveries like penicillin cannot happen twice. Progress is continuous, nevertheless. Ideas sprout, merge, and fade, seeding new branches of inquiry. What looks like a small offshoot today can grow into a major field decades later. It is true that most "easy" discoveries have already been made. All innovation rests on a pyramid whose foundation is careful research that has taken decades to amass.

But to say that science is no longer transformative is nonsense. Look at CRISPR-Cas9 gene editing, targeted cancer drugs, mRNA vaccines, GLP-1 agonists, antiretroviral drugs that have made chronic HIV infections manageable, artificial intelligence enabled by large language models, carbon-fiber aircraft, satellites powered by ultraefficient solar panels, and Google search. Contrarians take the position that the stagnant or declining life expectancy in America is the result of a failure of science. It is not. There is only so much a researcher in a lab can do to reduce levels of obesity, alcoholism, drug use, car crashes, violence, and suicide. And it takes time to achieve science. It also takes money.

Chapter 10

Much of what we have come to expect about grants, funding, and the financial system that keeps higher education upright changed in the initial days of the second term of President Donald Trump. Scientists and program managers at federal agencies were downsized, overhead rates on grants became the subject of legal fights, and research funds from some of America's most prominent universities have been withheld. Much of the targeting of universities arises from the pretext that they are promulgating leftist ideology.

One of my grants from the NSF—a collaboration with the University of Texas MD Anderson Cancer Center—was among 3,500 projects flagged by Senator Ted Cruz's committee as "woke" or "Marxist." The work concerned a device to help throat cancer patients recover more quickly. The offense, I suppose, lay in my biography, which mentioned my role as faculty director of UC San Diego's IDEA Engineering Student Center—an acronym for inclusion, diversity, excellence, and achievement—and in a small allocation to IDEA programs.

My involvement with the IDEA Center has been one of the most rewarding parts of my career. By the end of my term, it supported thirty programs that reached 6,800 students, funded scholarships and travel, and managed ten thousand square feet of engineering project space. Participation was open to all, though outreach was concentrated on students from low-income backgrounds. Because family income correlates with race and geography, our programs naturally drew from both minority urban and rural white communities. Favoritism based on race or sex has been prohibited in California since Proposition 209, passed by the voters in 1996.

The IDEA Center's success was built on the foundation laid by my predecessor and mentor, Professor Olivia Graeve, an immigrant from Tijuana who emphasized hard work and accountability as paths to stability. Far from a caricatured "DEI bureaucrat," she is a materials scientist whose research serves aerospace and nuclear applications—and she was honored by President Trump's White House in 2020 for expanding access to STEM research.

Efforts like ours to broaden participation are not "wokeness" but an expression of the American instinct to support the underdog. In any case, research focused on underserved communities represents only a vanishingly small portion of the US research budget. For example, in my grant called out by Senator Cruz's committee, $11,100 of a $600,000 budget (less than 2 percent) was allocated to outreach activities. These activities included the creation of a highly successful mixer series to match students with partners in industry, a mentoring program to help students get admitted to graduate school, and the creation of my podcast, *IDEAs in STEM Ed*. All of these activities were, to reiterate, made available to all students.

There was a notion in the early 2000s that we were shedding the set of beliefs that Americans had so long held onto: that people could just pull themselves up by their bootstraps, irrespective of background. People were recognizing that race, financial status, or both were at least partial determinants of academic success. Universities and lawmakers were working to help those who had fewer advantages achieve success, because success in academia is well correlated to increased socioeconomic status. Students from low socioeconomic strata are less likely to have family members who know about careers in STEM fields, friends and classmates who are pursuing college, and awareness of financial aid programs. They are also more likely to experience financial stress, an aversion to taking out loans, and the expectation of giving money to their family. This isn't to say that true Horatio Alger stories don't exist, and it may be easier to come from nothing in the sciences than it is in other high-status disciplines. Indeed, most of my PhD students were from working-class backgrounds, and many were also the first in their families to go to college. Several were members of ethnic minorities to boot. However, to say that it is *possible* to pull oneself up by one's bootstraps does not mean it's *probable*.

Almost immediately, I was inserted into the diversity landscape at UC San Diego. In addition to the course that I inherited from a departing faculty member, I also inherited a position as the diversity officer within the department. Eventually I also inherited leadership of the IDEA Engineering Student Center. In 2017, because of my work within the center, I received the Diversity Champion Award on campus, the highest award given to those whose work supported women, minority, and low-income students. When I received that award, it should be said, there were few women working in the lab. I felt self-conscious receiving it. I had done a lot of outreach—at least one hundred events for the recruitment and retention of women and underrepresented minority students to engineering. But although I had many undergraduate women in my lab, I had few women PhD students and postdocs until later on.

One time I hosted a retreat for my lab group, renting a huge house in Big Bear. We did day hikes, a trivia competition, and other activities. During an evening at the house, when we were talking about the direction of the research group, one of the students expressed concern. "We're getting the reputation of having a bro culture," he said. There were rumors among a small number of female graduate students outside of my lab that I did not recruit women. This was pretty crushing to me. I grew up in a house of sisters. My mom was my intellectual mentor. My graduating class was populated by female achievers. My postdoc advisor was a woman.

I started to reflect. Was it me? Was there something wrong with my hiring practices? Was I facilitating a toxic culture within the group? I was in a field that was dominated by men: Four of five chemical engineers and materials scientists are male. But the perception didn't match what was actually happening, as the tide was already starting to turn. Laure had already joined, and B, Allison, Rachel, and Laura would be joining soon. Moreover, our undergraduate cohort had a higher percentage of women (nearly 40 percent) than the department average (20 percent)—which made the perception of a "bro culture" all the more frustrating.

The composition of the research team matters, but not only for demographic or cultural reasons. A lab is a workshop for ideas, and the point, in an ideal world, is to create something that can make its way into the world. The people you recruit shape what is imagined, what is attempted, and what survives contact with reality. During my time in San Diego, my students had their share of projects that went nowhere, but also some that worked. One of the biggest rewards of running a lab is nurturing creative and hard-working individuals, watching them take their project further than I could have, and then seeing them go out into the world and do something with it.

One of my most productive early graduate students was a helpful team member but was highly independent. I felt that he humored me by working in areas that were core to the group. Nevertheless, he really wanted to devise and execute his own projects on a type of material that was all the rage in the aughts through the twenty-teens. The material is potentially useful for many things, including biosensors, solar cells, computer monitors, smartphone screens, and tablets.

I could tell that the student was a catch: brilliant and hardworking, with a forcible personality. He didn't arrive in my group intending to be an entrepreneur but stumbled upon a possible business opportunity for the material he was making. The material was really expensive—hundreds of dollars for a square inch of it as a thin sheet. Additionally, the quality control of the commercial suppliers at the time was dismal. Sometimes you would shell out hundreds of dollars and get a highly defective sample or a blank substrate with no material on it at all! It was super thin and mostly transparent, so flaws were easy to miss during visual inspection by quality assurance personnel. My student devised a method to make lots of this high-quality material at a lower cost.

Since my research funding was from grant proposals and through my startup from UC San Diego—already committed to other projects—my goal was to find an alternative source of funding for him. We found a handful of small business pitch competitions, and I helped him craft a business plan and a slide deck. I felt like my business training from Stanford was coming in clutch. It turned out that he was also an excellent public speaker, and his project earned a $50,000 award from the Department of Energy. He used these funds to pay a law firm's fees to write a patent application, file the paperwork to register the company, and reimburse my lab for the purchase of the equipment he had asked me to order to demonstrate his process.

My student's success in earning his own funding was remarkable: He brought the company into existence by charisma and force of will. Since he raised his own startup funds, the university didn't end up taking an equity stake in the company, although UC San Diego did "own" the patent. This last point is remarkable. Usually universities pay for patenting costs because the startup formed by the students has no money. In exchange for taking on all the risk in case the startup fails, the university takes equity in the company through its licensing agreement. For example, when Larry Page and Sergei Brin took Google public, Stanford netted over $330 million when it sold its shares.

Many universities have entrepreneurship centers, which—nominally—help students take ideas from lab to market. These centers attract a gaggle of know-it-alls: local businesspeople who give advice to would-be student founders. Such individuals project extraordinary confidence. In this context, my student came across a successful businessperson who made money in a large company. He then became a startup founder and professional angel investor. He was so enthralled by my student's vision that he gave up his leadership of the angel investor network and became the full-time CEO of the nascent startup.

Usually, when a startup spins out of a professor's lab, the professor takes on an advisory role in the company and becomes a shareholder. In some cases the professor takes on a much larger role, such as chief technical officer or chief executive officer, necessitating a leave from the university, but this is rare. Professors are not hired for their business acumen and are not known for being dealmakers. What is uncommon is for a professor to have zero involvement. After all, the professor was presumably interested in and invested in the research and could be valuable to the company in the future: making connections with customers and partners, recruiting employees, and keeping the company abreast of developments in the field. In my case, I had already invested my time and laboratory startup funds in research materials.

My student, however, was initially skeptical of having me involved. He thought I didn't want to do the science that he wanted to do, that I was resentful of the work he was doing. His work was in the general ballpark of my research, but I had no experience with its methods, and it showed. Even so, I continued to have regular meetings with him, supported his research financially, and directed potential collaborators and investors his way. Nevertheless, the question of equity took some time to resolve. Who got which shares of the company?

I got advice from our tech transfer agent at UC San Diego, Chip, who was a mentor to me during this period. He told me that I had come in too hot the first time. It was true—I had lost sleep every night for weeks with distracting, intrusive thoughts on repeat. I can see now that I shouldn't have been looking at it in an adversarial way. It wasn't about what I had done for my students in the past; it was about providing them with the resources they needed to flourish in the future.

The new CEO asked me to go for coffee one day. I was, he noted, creating issues. I was regarded as a loose cannon. They worried that I was going to interfere with patent licensing. Any new private investors were going to do due diligence, and I needed to be on the startup's side. I wanted equity in the company, and I felt I deserved it. We met up at the outdoor café at the Whole Foods in La Jolla.

"It was his idea, his brainchild," the CEO told me. "He's a genius."

"He's extremely smart, I'm not disagreeing with that," I said. "But I also helped him craft his elevator pitch before you were ever involved. I critiqued his slides. I fronted him the money for the instrument he installed in my lab that made his device. My lab purchased all his materials and underwrote his facilities usage. And I can help you, because I'm connected to students with the skills you need in this type of business. I can help you find customers. I know chemical suppliers and many people in academia doing this kind of work."

After I enumerated what I could offer as a scientific advisor, my student and the CEO wrote up an agreement. They offered stock options. They rented a space in San Diego and started cranking out material for the research and development market, selling it to other researchers and companies and distributors that would provide it and sell it at a markup. They had a smart business strategy, which was to generate income from a commoditized product. In the background, they searched for bigger opportunities in biomedical devices and microelectronics. While my involvement has never been intensive, I helped them recruit personnel, spoke on their behalf to investors, connected them to customers and vendors, and brought them additional funding through a grant. Over the last ten years, my relationship with the company and its founders has strengthened. When my stock options vested, I purchased them.

Growing pains in a startup company are inevitable, whether the company emerges from a university or not. Challenges notwithstanding, projects incubated in my lab or started by my former students have been successful. They are still in business, growing, and their founders are winning accolades. They're examples of how incredible people and ideas can be nurtured in the setting of an academic lab and go on to influence the greater good when the federal world combines with the educational one. Estimates place the knowledge created by federally funded academic laboratories as responsible for 25 percent of economic growth in the United States since World War II. Already impressive, this statistic omits the creation of human capital in the form of over forty thousand PhD holders in science and engineering—nearly all of whom were trained on projects funded by grants underwritten by US taxpayers.

I take equal pride in the many former students who have gone on to build impactful and secure careers in established industries. Rory, for instance, took a job advancing solar technologies at a large company, helping design roof-integrated photovoltaic systems. Cody combined his technical training with an MBA and now bridges science and business at Eli Lilly. B works at Illumina, the world leader in DNA sequencing. Julian is engineering medical devices at a company that has transformed the lives of diabetics worldwide through its real-time glucose monitoring. Brandon, a decorated inventor at Intel, contributes to the resurgence of US semiconductor manufacturing. Daniel develops next-generation equipment for fabricating computer chips, while Don Garrett contributed to the development of electric cars at Tesla and later Rivian. Rachel and Laura are working in smaller firms, making innovative products at the intersection of biomedicine and electronics. Taken together, these careers show how publicly funded research and training pay dividends far beyond academia.

Chapter 11

Dina and I did not want to have kids. That was something we both agreed on. For the longest time, we were an anti-kid couple. The couple who gladly went out to dinner and enjoyed the silence. Who traveled the world without a second thought about babysitters or car seats. Who lived our lives for ourselves.

We were annoyed by kids on airplanes. We were annoyed by kids' helplessness. We joked that babies had faces that looked like heirloom tomatoes—scrunched up and red. We said we didn't want to have a tomato-faced baby.

When you're married for a time, you and your relationship start to evolve. You start to feel different. There was no single moment when Dina and I looked at each other and said, "OK, now we're ready for a tomato-faced baby." It was never quite that clear. Eventually, though, we came to understand that our love could support more than just the two of us. I also saw that if my sister, Deena, the co-inheritor of the Lipomi gene for anxiety and the tendency to catastrophize, could successfully become a parent, then so could I. Plus, I enjoyed hanging out with my little niece, Mariana—named after our father. Mari, for short, is a cool kid; she filled our family with joy during the holidays, when we all reunited at my parents' house in Hilton.

In my marriage, I realized that I was not the easiest person to live with. And Dina knew she was not the easiest person to live with. I'm an anxious person. I'm self-critical. Dina is a lot like me in this way. As scientists, we are in the camp of biological determinism, and we were worried about transmitting genes for anxiety and depression to our child. As environmentalists and overthinking millennials, we were also concerned about the drain of one more mouth to feed on our overburdened planet. On the other hand, what if we had a child who one day invented a type of solar panel that was 100 percent efficient and could be purchased for a dollar? At one point we decided that there was no rational reason either to have or not to have a child. Internet commenters and editorialists couldn't see the future any better than Dina or I could. So, we started the journey.

Dina's ob-gyn had informed her that it would be difficult to conceive. We were told it would be a long shot, but by some miracle she got pregnant anyway. I have no idea how it happened. At the time we were intense cyclists. We did exactly what they tell you a woman is not supposed to do in the first weeks of the first trimester. We were on a fifty-four-mile round-trip bike ride from San Diego to Carlsbad and back. We would often do a ride like this on both Saturdays and Sundays. When we got back, I was reading one of the Game of Thrones novels. We were four months into trying to conceive. I thought it wasn't going to happen. Dina thought it wasn't going to happen. Or maybe we thought it was going to require some kind of medical intervention.

"The pregnancy test came out positive," she said.

I remember sitting on the patio, looking up from my book, adjusting my glasses, and saying, "Quite."

Dina, who was professionally well versed in the potential for error in at-home diagnostic tests, went to Rite Aid and purchased a test from another manufacturer. "That test came out positive, too," she told me.

Soon thereafter, she went to the doctor and got a blood test, and it was confirmed: She was pregnant.

We should have been over the moon with joy, but I remember being worried too. Because of the ways in which our lives would be transformed, sure, and also because of the possible teratogenic effects of the recently completed interior renovation of our home. There was smelly glue and new paint everywhere. We had been doing things— cycling long distances, naturally—that the internet had told us not to do. Had we already failed as parents?

In every ultrasound we ever got of Cora, our daughter, she was breech, meaning that her head was pointed up. She was stubborn and never flipped. We tried to change her position using an external cephalic inversion procedure—during which they hook the mother up to all the equipment that she would be hooked up to in the event that the procedure accidentally initiated labor—and that was a failure. So Cora was born by C-section.

Nothing can prepare you for what happens when your child is born. What shocked me most about welcoming my child into the world was that there was already a personality in Cora's face. This was Cora, who was already thinking things, who already had emotions and predilections. At the beginning of having a newborn at home, I thought it would be a little like raising kittens, which I had done before. They can walk and find food within a month. But no. A baby is totally useless until they're two.

Most nights it's hard to get Cora to sleep, and this was true even at the very beginning. She sleeps through the night, and she has slept through the night ever since she was three months old, which was amazing. Getting her to sleep to begin with? Seven years in and still a source of frustration. Like her mom, she is a night owl and doesn't seem to have an off switch. Nevertheless, I often long for those early days when she would just drink from a bottle and sleep on my chest in a recliner while I listened to music or read the news or typed an email with one hand.

We had a great neighborhood in San Diego. We lived in the first house on a small road that ended in a cul-de-sac, in a neighborhood where the houses dated from the early 1970s. We're still in contact with the people in our old neighborhood. In a way it felt like how my father described his old neighborhood in downtown Rochester. We would watch each other's kids, celebrate birthdays, have block parties and Easter egg hunts, and throw back a few, unrestrained by the need to drive, since we could just walk home. Work friendships were different, kind of contingent—contingent on physical proximity and shared jargon, if not always shared purpose. There is often an undercurrent of competition among others in your cohort, especially in the world of university research. In fact, the system bakes it in: You get promoted based on explicit comparison with others at your career stage. It's perhaps not surprising that most of my enduring friendships from my twelve years at UC San Diego are with my former trainees.

We tried to have another child, but lightning did not strike twice. We tried for six months, and then life got really stressful during COVID. We decided to stop trying. Cora, whom I love with my whole heart, is a challenge. She does things that I know I didn't do as a kid: For one thing, she loves swimming, which I didn't learn to do until I was twenty-five. She is also high energy and talks back to us a lot, which is possibly a good sign for her future confidence but which can be frustrating now. Dina and I had names picked out for future children—Evan and Zoe—but we gave them to our black-and-white tuxedo cats.

In 2020 our family trip to Hawaii was canceled due to COVID and was rescheduled for 2021. The state of Hawaii required us to have our COVID vaccines. Despite the threat of the virus, Cora was still going to daycare. It was a home daycare four houses down the street from us in San Diego, run by a beautiful older couple who spoke Spanish in the house they'd owned for nearly fifty years. Some of their helpers crossed the border on a daily or weekly basis to come to work, and they all loved Cora. Then the couple and several of the kids got COVID. They had to close the daycare for ten days.

We knew there was *a chance* Cora had been exposed, but we were asymptomatic, and we all tested negative using the most accurate form of molecular test, so we took our chances and proceeded with our travel plans. We arrived in Hawaii, and everything was fine—until it wasn't.

Cora, age two, started feeling lethargic and had a low-grade fever. There was no logical explanation other than COVID, but we were in denial: *It's just a cold.* It was a redux of my experience in Peru, where I opted to believe that my incipient sickle cell episode was just indigestion, until it became impossible to ignore. She lost interest in eating, even candy. The color in her face, ordinarily kissed by the summer San Diego sun, disappeared. We put a mask on her while we surreptitiously carted an obviously sick child through the airport. We got her back home. This was in the middle of the Delta variant, which had the reputation for causing breakthrough infections in the vaccinated. I had been quadruple-vaccinated, because I was part of the clinical trial for the AstraZeneca vaccine. As an education worker, I was deemed "essential" and had early access to the Pfizer vaccine. I felt bulletproof, like the Terminator.

Despite my molecular armor, I too began to struggle with my breathing. It was like having a heavy weight on my chest. Cora was like a rag doll. And we were led to believe that children were at the least risk of developing bad COVID symptoms. The doctors told us not to bring her in unless it was truly serious—but how can you tell, as a parent, what the threshold is between *serious* and *life-threatening* when you're talking about a two-year-old? When her fever started to break, I felt an enormous sense of relief and gratitude. It would still take her several days to recover, but arriving at the inflection point—improvement of symptoms—was crucial to my psyche. While I was having trouble breathing, which at times brought me to panic, my symptoms felt as though they instantly dissolved once I was sure Cora was going to be OK.

Looking back, I now see this as a formative moment in my parenting, when I started to understand the depth of my love, the depth of my advocacy, both how strong I was and how helpless. How do you care for another person? How do you make sure that you are making all the right choices? You never can, of course. You do your best. Parenthood brings out in us our fiercest love and loyalty, our starkest defenses. Cora was also now a member of the club: the third member of our family to fight through a life-threatening event while on vacation.

Chapter 12

In 2016 I received the Hall of Fame Award from Hilton High School. I was deeply honored to be invited back to my school and to be in the Hilton Homecoming Parade. It was a full day of activities, and I was scheduled to speak to some of the chemistry and physics classes. Afterward, I would attend a scheduled lunch with faculty and administrators. Then I would be part of the homecoming pep rally in the gym—a far cry from my own time at Hilton High School. As someone who had been a complete and total nerd in school, who had suffered every sling and arrow thrown by the jocks, I was now all the way on the other side. Cheerleaders and football players were doing all the things they had always done, of course, but somehow I was in the center of all of it.

Right after the pep rally, I was going to be featured in the homecoming parade and do the coin toss at the football game. It turns out that I would do neither. I received a call from my mom.

"Your father is OK," she said. But what followed was anything but OK. My parents were on their way to a routine visit with my father's primary doctor, whom he disliked. My father had become increasingly agitated in the previous several months but became downright aggressive in the car. He grabbed my mother's wrist. When they got to the clinic, my mother got the doctor in a corner and asked for help. My father was off the rails. The police came. They strapped him to a gurney and took him in an ambulance to the emergency room.

My father was eighty-five at the time. He had been showing signs of dementia starting five years earlier, but had not reached the level of crisis until his breakdown at the doctor's office. He was transferred to the University of Rochester's Strong Memorial Hospital, where he remained for more than three weeks while they figured out how they could safely return him home.

In the meantime I was living my best life, traveling like a VIP and living like a big shot within the scientific community. In fact, I was supposed to fly right from the event in my hometown to Jeju Island in South Korea to attend an international conference on polymers. I arrived at the emergency department in a suit and tie, still dressed for the homecoming event. I felt ridiculous, like I was playing dress-up in the most serious of places.

Maybe it was just the gut instinct to flee a place that had made me so uncomfortable, but I went anyway. I flew to South Korea. I knew that my father was in the best hospital in the region. I felt useless in Rochester and useful in Jeju, so that's where I went: somewhere that I could serve a purpose. To my credit, I talked to my father on the phone every day of my trip, trying to distract him with stories.

At the hospital, the doctors took CT scans of my father's brain, which revealed advanced dementia: There were black parts in his brain, appearing like deep crevices, indicating reduced blood flow. My father's brain resembled a shrunken raisin. The geriatric psychiatrist showed the images to my father, who did not react.

The doctors prescribed my father Seroquel, which was ineffective, and then risperidone—a magical cure. For five years, his aggression mostly disappeared. He still suffered from some paranoia, and he lost his storied facility with arithmetic. My father had been given five extra years and a chance to have a short relationship with his granddaughters. He had been turned back into his old self, for the most part. It was unbelievable the salvation offered by this pharmaceutical intervention.

Cora was born a year before COVID, and a whole new life was starting for our family. Because of this, and of course COVID, I was no longer able to make my quarterly visits to Hilton to visit my father. I knew I was missing out on the last few years of his life, though we spoke by video chat often. He liked to hear about little Cora's baby steps:

eating, walking, and talking. My mom, seventy-six in 2020, finally retired from her job. Right before the pandemic, she had finished a second master's degree, in library science, and she was working at the library as an interim director. But she was coming into contact with too many people and didn't want to bring the virus home to my father.

In the summer of 2021, right after our trip to Hawaii and our own battle with COVID, during the height of the Delta variant of the coronavirus, my father stumbled and fell in the bathroom. We knew he had congestive heart failure. He would often get lightheaded and experience sudden bouts of weakness. My mother was there to prop him up until the ambulance arrived and took him to the hospital. For the five preceding years, since his initial meltdown at his doctor's office, he had been on the brink of being put into a nursing home. It wasn't an outcome we wanted. For one thing, we never knew how we would pay for it. My mother had worked into her late seventies to pay off her mortgage, and she didn't want to lose her house. My mother, sisters, and I had many older relatives and spent a lot of time in nursing homes. The thought of my father wasting away in a crypt was something we wanted to avoid at all costs. As long as my mother could get him to the bathroom, she said, or into his chair, she was confident she could care for him at home.

Then, one morning in August, she received a call from the rehab center in a wing of a hospital. My father, they said, was in critical condition and unconscious.

My mom called me. She was crying. I told her to go to her neighbor Sally's house. It was early in the morning in San Diego.

She called me a second time. "He's gone. Can you call your sisters?"

Maybe she called me first because, even though I am the youngest, she believed I had the evenest keel. I called my sister Andrea in Las Vegas. I called my sister Deena in Rochester. I went upstairs. I was sobbing. I told Dina what had happened.

I bought red-eye flights for Andrea and me. For me, from San Diego, and for her, from Las Vegas.

Arriving in Rochester groggy and barely functional, my family and I made funeral arrangements with a family friend, Tom, who owned the one funeral home in town. This was the big advantage of being in the small town of Hilton, because everyone knows everyone. Tom had buried most of our relatives and took over all of the arrangements for the funeral and burial. I was so grateful for the burden he bore for us.

We did a showing for the immediate family. No one can prepare you for the profound, deeply emotional experience of losing a parent, of seeing my father's body with the cold packs inserted under his best suit and with the peaceful expression fixed by the funeral director.

We honored my father's life at St. Leo's, our family's long-time church, with five hymns chosen by my mom and the church organist. I had heard them pretty much every week for seventeen years. Hearing them again—"Shepherd Me, O God" and "On Eagle's Wings"—was revelatory. Hearing them at any time since has been as sure a way as any to bring me to tears.

Due to the threat of COVID to older people, my mother, approaching eighty, wanted to keep things small. We may have offended some friends and relatives by inviting only the closest relatives. There were no calling hours, just the funeral mass. I delivered his first eulogy. In it I acknowledged that my father's age may have embarrassed me as a teenager, but as an adult I took it as a point of pride. "My dad is eighty!" "My dad is ninety!" I spoke of my father's prowess in card games, having beaten him in gin rummy only once. I recalled our summer nights watching baseball on TV. Despite his conservatism, he always supported the underdog, which meant anyone playing against the New York Yankees. How he taught me to drive. Once I got my license, he would take me to car dealerships to take test drives of used cars we couldn't afford. "No one is better than us just because they drive a fancy car," he would say.

"Don't criticize people for things they can't change" was my father's golden rule, and Dina and I have adopted it in our own family. I concluded by acknowledging my mother for the many sacrifices she made during my father's declining years. I remember speaking these words at the lectern, adjacent to the altar, in a mostly empty church. The reverberations in the room seemed to add significance, even when I barely squeaked out, "Dad, we love you," in conclusion as my voice gave way to tears.

My sister Andrea delivered the second and final eulogy. She spoke beautifully about his legendary hosting. Family parties at our house in Hilton. How he always wanted people to feel good in his presence. How he had a talent for meeting caring people and adopting them into our family. How our father was a protector and servant to those in need. How he loved his family. Despite the dysfunction and acrimony of her teenage years, it was clear that my sister loved him. "Each day with you was a gift. And I'm happy that you can rest now," she said as tears welled up in her eyes and those of all of our small gathering.

My dad had always said that he did not want to be cremated, because he was Catholic, but we did it anyway. It's easier, much less expensive, and felt—having seen crypts and full-body relics with mouths agape on my travels—somehow more dignified. We buried him in Holy Sepulchre Cemetery in Rochester, where my mother had buried both her parents. It occurred to me that Mom had stood in that exact same spot for her dad when he died in 1960, when she was just a teenager. She stood there again to bury her mother in 2001, not long after 9/11. Two decades later, there to bury her husband. Her whole life had been there, love and life stitched together in this one square yard of earth.

I appreciate things about my father much more now that he's gone than I did when he was alive. As the youngest and only son, he took care of his mother. Because of my family's age disparities, his nieces and nephews are more like my uncles and aunts, even though they are cousins. They'll tell me stories about him as a younger guy, about how he drove them around and got them treats, and that he was the cool uncle—they called him "Uncle Moni"—who everybody loved.

So there are the facts about my father. He was an old dad who found love late in life. He cared deeply about people. The things that felt embarrassing as a teenager have become the things I hold closest: his stubborn loyalty to people, his emphasis on dignity over status, the ease with which he made others feel seen. He wasn't a philosopher, but he left principles to live by. He was not an intellectual, but he was curious. He was then, as he is now, larger than life. Mariano. Uncle Moni. Mars. Dad.

Chapter 13

The loss of my father was, of course, monumental, but it was fair to say that things in my life were still on the upswing. Cora was thriving. I was happy in San Diego. The weather suited me: I maintain that the best outdoor run in America is Black's Beach to Torrey Pines State Park at low tide. It's possible I did that run five hundred times in the twelve years I lived there. I would do the run barefoot and stop at the turnaround point to practice my version of mindfulness meditation (so Californian!). The beach was so close to my office that I would often abscond during the day to run there, then return to work, spiritually cleansed and ready for anything. My work suited me. I had a great home and a family—maybe one that I had never even expected. And then, much to my surprise, an opportunity arose that I couldn't ignore.

From 2016 to 2018 I served as the director of the Graduate and Postdoc Scholarly Talk series, called Grad Talks, at UC San Diego. My role was to select the topic—for example, writing, speaking, and time management—and to invite the speaker. When I couldn't find someone to give a seminar on the topic, or if we had a cancellation, I would give it myself.

I ended up with a cache of eighteen talks during those first few years, which I put online, and they became the basis of my YouTube channel and my podcast, *Molecular Podcasting*. They ended up collecting a lot of views, some videos in the tens of thousands. As a result of that work, I was invited to become the faculty director of the IDEA Center. A year into the job, it was merged with the associate dean for students for the School of Engineering. I took on that role too.

I began to view my work with video production as a continuation of my childhood interest, starting with the PXL2000. From 2017 to 2024 I produced over two hundred videos on my YouTube channel, which garnered millions of views. Because of this profile—and also because I had the word *dean* in my title—I started receiving emails from headhunters for leadership roles such as department head and even dean at other schools.

The notion of being poached by other universities certainly boosted my ego. Was I a hot commodity? Could I, in fact, go elsewhere? Maybe the plan I had made for myself—a high-status job in one of America's most desirable places—wasn't set in stone. I started to explore these external inquiries with increased intensity. I interviewed on Zoom for a job heading a famous chemistry department in a rural location to which I knew Dina would never want to move.

Per Dina's suggestions, I started looking for positions that made sense for both of us. Like before, I considered Austin, but then another thought occurred to me. What about Rochester? What about going home again? Rochester made sense for several reasons, not the least of which was that it was close to my mother, my sister and her family, and my cousins. It's also beautiful (most of the year, anyway); has a good food, beer, and music scene; and has sandy beaches (believe it or not). It also has low housing costs relative to the area's beauty and amenities because the snow scares most people away.

For her part, Dina needed the right ecosystem to take advantage of her experience in both science and business. She was a virologist by training and had earned an MBA from UC San Diego in 2018. Leveraging her scientific and business training, she worked at startups (pre-seed companies literally operating in garage labs) and venture-backed companies, not to mention a stint at a large Fortune 500 company. Finally, she successfully ran the San Diego site of the largest biotech incubator network in the United States. She had a very specific idea of what she wanted to do and where she could add value. By 2023 she considered herself more of a businessperson than a scientist.

Rochester is certainly not Boston or the Bay Area, but the University of Rochester is the largest private employer in the state outside of New York City, and its prestigious medical center conducts NIH-sponsored research in the hundreds of millions. So there is—and remains—opportunity for innovation and startups.

I never thought I would be going back to Rochester, but life, as I see it, is a series of accidents. I think we have some idea that we are fulfilling some movie-like arc and that we are supposed to make some decisions based on predilections and interests. Thus, you can never really predict what's going to happen.

Every career move starts with a push but ends with a pull. Here was the push—though perhaps "nudge" is more accurate: UC San Diego had been remarkably good to me. The University and the Department of NanoEngineering believed in me when I had no other offers. They gave us the down payment assistance and mortgage program that let us buy our first home. They subsidized Dina's MBA with a huge scholarship. I'd built a research program there, raised my daughter in paradise, and run barefoot on Black's Beach hundreds of times. I was genuinely happy.

But by 2023, small frictions had accumulated in ways that made me restless. Being in an administrative role exposed me to truths I couldn't unsee. How the red carpet can be extended for faculty members who are unkind to their students and staff because they bring in grant dollars and publish in fancy journals. How money and efficiency sometimes trump morale. For example, the campus installed sensors outside offices and labs to track occupancy, justified as a way of measuring the utilization of space but experienced by the faculty and students as creepy. In an unrelated incident, the graduate students and postdocs went on strike. While the strike was ultimately resolved with higher pay, it still revealed tensions I hadn't fully appreciated. For example, the administration seemed to believe that students' identities as both learners and workers could be neatly separated. Moreover, some PIs gave grades of "unsatisfactory" for research progress, despite the legality of the strike, as if one could pause being a researcher while continuing as a student.

Then came an incident with a faculty candidate my department wanted to hire: We had identified a brilliant young scientist and negotiated their startup package, and they had resigned from their job.

Too soon, it turns out, as they never had the official offer letter from campus. Our provost canceled the search because we failed to hit our enrollment targets for our master's program. Watching the candidate process this betrayal was simultaneously heartbreaking and humiliating, as I had reason to believe that everything was on track, and told them so.

None of these items individually was catastrophic and could have happened in any university, or any large organization full of well-meaning people driven by competing mandates. I wasn't angry at UC San Diego—I was just ready to listen when opportunity knocked from an unexpected direction. Sometimes you don't leave a place because it's bad. Rather, you leave because you've become a different person than the one who decided to be there.

I loved the beach. I loved being outside in the sun in January and February and taking a cold plunge in my own pool in March. Many of the people who learned that I was moving from San Diego to Rochester questioned my sanity. "You're applying to Rochester?" they quipped. "Are you insane?" And, I'll admit, sometimes I asked this of myself, particularly when I thought back on all the brutal winters that I had been running from. I remembered, for instance—and with great clarity—the epic ice storm of 1991, when we lost power for a full week. Others in the region lost it for much longer. Why on earth would a person subject themself to this transition on purpose?

But happiness, I've found, equals reality minus expectations. Happiness, too, is not a straight or conclusive path. There is no one way to get where you're trying to go in this life.

Then came the pull. There was a lot about the position at the University of Rochester that appealed to me. I had found it on the department's website. They were hiring a mid-career or senior faculty member to take on the role of department chair and establish a strong research program. The salary was good, and although the department had been flagging during the early 2000s, that provided me with an opportunity.

When I agreed to check out the program, it was right before Christmas in 2023. I submitted my application, and the department chair reached out shortly thereafter.

They appeared skeptical that I was going to come. "This sounds like a downgrade," one of the faculty members told me. I had $10 million in research expenditures over twelve years as a professor at UC San Diego and could theoretically go many other places, while the chemical engineering department at Rochester was in a state of rebuilding. Six people had been hired in the previous two years, and I was the only one who came in at the mid-career level. The other five? All brand-new. All but one—who started her career at Barnard College at Columbia—came in fresh out of their postdocs.

But what I saw was a real opportunity. I could build a program from the ground up. I could help the program grow. Because of where the University of Rochester is located, it was historically hugely influential in the development of Kodak and Xerox and Bausch & Lomb. The program deserves to be regarded with the prestige that is commensurate with its history and contributions. I began to reframe in my mind what opportunity lay before me.

The decision of whom to select as your PhD advisor and whether to marry, have kids, or move to a new job are what economist Russ Roberts called "wild problems" in his eponymous book. The person you become after having made such an affecting decision is different from the person who made it. For example, in the decision to have a child, having fretted over the loss of personal time seems childish in comparison to the privilege of getting to raise Cora to adulthood. Similarly, having a real option to change the circumstances of one's life and one's family— financial security, professional agency, and reuniting with family—ended up outweighing concerns I had about giving up our pool. On the level of ego, it helped that I was being hired not only as a tenured professor but as department chair, and that the position came with an increase in salary. In terms of academic hierarchy, associate dean outranks department chair. However, the role of the chair is as an executive, whereas that of the associate dean is to do the dean's bidding. The chair is the captain of a starship, whereas the associate dean is the rear admiral doing paperwork on Earth.

My research in the past two years has been in a period of reorientation. It has been helpful, in some ways, that the school did not immediately have a lab for me to move into. They are building my lab at the time of this writing, and I am using a temporary space, which is working out just fine. I don't know right now if I want or need to crank out papers at the same rate I achieved at UC San Diego. A new place comes with new opportunities, and I've been aligning my research interests with the unique strengths of U of R in optics, neuroscience, and music. I've been enjoying the side of my job as department chair that encourages me to develop relationships. I've been getting to know the alumni and local companies. I've been enjoying returning to my roots too. The Eastman School of Music is right here. As a high school student, I wanted to go to the conservatory, but I didn't get that opportunity, and now I have it at my fingertips. I picked up the trombone again and have a small collection of rare and vintage instruments. I have gotten to know Larry Zalkind, virtuosic player and professor of trombone at Eastman, who now occupies the studio in which I so nervously flunked my audition a quarter century earlier. Together, we are writing a book on the physics and physiology of playing brass instruments.

It's incredibly fun to be able to work on the humanistic side of engineering, which is what I have wanted to do ever since my days reading E. O. Wilson, back in my freshman year of college. I'm enjoying the fact that U of R is not nearly as focused on superficial metrics: papers per year, dollars generated, and butts in seats. We deliver a high-quality product in the classroom, and we don't do things to game student evaluations. This is a hard school. It's expensive. I feel a bit of guilt about that, about leaving such a huge program that brings so many people out of low-income backgrounds. But we do that too. Most of our domestic students aren't paying full tuition. In my new role, I've also kept my

YouTube channel going, and its audience—now at twenty thousand subscribers—is growing.

Much of my work revolves around fundraising for the department, engaging with alumni, getting speakers to appear on campus, increasing opportunities for students and faculty, and generally amplifying awareness of our program. With the support of a small group of alumni, in particular a large donation from trustee and alumna Barbara J. Burger, former president of Chevron Technology Ventures, I created a new scholarship in chemical sustainability. This program is modeled after the Beckman Scholars Program and the concept of learning-aligned employment—that is, getting paid to work in a university research lab without having to take on a work-study job that is unrelated to one's major.

My quality of life has also improved since I returned to Rochester. The house in San Diego that Dina and I bought in 2013 appreciated in value by nearly a factor of three. That good fortune, paired with UC San Diego's mortgage program, allowed us to buy our home in Rochester in cash, which was less than half as expensive as our home in San Diego. (In part because we had such favorable mortgage terms, Dina and I were able to make a pledge to support the IDEA Center at UC San Diego upon our departure.) Our home here is on an acre of land. There is a creek in our backyard, through which deer, groundhogs, and foxes saunter. Our wooded yard is an aviary: Cardinals, bluejays, woodpeckers, robins, and chickadees are ubiquitous.

The region is full of industrious and caring people, thriving businesses, and myriad cultural opportunities. Sure, there is a lot of snow in the winter, but it somehow makes the coffee, tea, and whiskey taste better. The landscape is green for seven months and white for five: all evidence of an abundance of water, which we did not have in SoCal.

These days, Cora has a life that I always dreamed of for her too. In returning to Rochester, I've reunited my family. My mother is here. Deena and her family never left the area. Cora gets to see her first cousin, Mari, as much as she wants. Shortly after we moved to Rochester,

Andrea, along with her partner Paul, felt the draw of family, and moved from Las Vegas to Rochester. We have reunited with my cousins—much older than me because of the age of my parents—their children, and their children's children. We have a family, large and unruly and filled with unconditional love and support. What is possible for us here, in New York, was not possible back in California, no matter how much I wanted to make that life work. We loved it out West, but there were parts of me that were always here, back East.

Dina works as the technology development program director in the University of Rochester technology transfer office, where she gets to build and grow programs for entrepreneurial researchers. Now, in Rochester, in the middle of my career, I am building out the chemical engineering program as chair of the newly renamed Department of Chemical and Sustainability Engineering. The name change was something I pushed for, and it reflects what I believe the field can be. This turn of events was not something I expected, but in retrospect, the decision to make this move couldn't have been clearer.

Epilogue

I have learned plenty of lessons in my time in higher education, not the least of which is the role of luck: Being born to caring parents in a town with good schools. Being born into a family that didn't enjoy the resources I have now. Yes, being white and male. Having access to modern ways of managing my clinical anxiety. Having the curiosity and freedom to pursue a set of interests, and ones that society finds valuable.

Not only have I held esteemed positions where I have been valued, but I have also been well compensated. I have been able to take advantage of incentive programs that have allowed me to purchase property. My quality of life has been connected to my job, and I have lived a good life. I don't deny my own good fortune, nor do I look down upon it. This career has been a good one.

When I started thinking about this book, it was only partially an exercise in self-reflection—it was an act of self-defense. In 2025, as research budgets were gutted, federal agencies dissolved, and scientists dismissed as political enemies, I realized how fragile the economy of discovery had become. For decades researchers were trusted to use a small outlay to advance science for the public good, but that trust is evaporating. The book tacked in a different direction as indignation turned into resolve. I reasoned that if I could pull back the curtain for the voting public and for policymakers on how discovery happens, maybe I could show what's at stake—university research as the invisible ant colony that lets your garden flourish and asks for little in return.

I had once imagined writing an exposé—a dispatch from inside the machine. Anthony Bourdain's *Kitchen Confidential*, but for research labs. But the longer I worked, the less interest I had in indictment. I found myself writing instead about the students, postdocs, and professors who make the system work despite itself. The world I know is full of those people, and I wanted readers to see them not as victims of bureaucracy but as participants in a fragile, ongoing miracle that has fueled economic development in America for eighty years. If this miracle is to endure, it can't remain a private conviction shared among practitioners. It must be visible, legible, and defended in public.

If it fails to be, it is our fault, as scientists. For decades, taxpayers have trusted—or tolerated—the use of public funds on basic research. The promise was that it would lead to discoveries for the public good. With this arrangement, scientists have been content to live in our own world, speaking our own language, working on projects that we love, and happy to be left alone. To reiterate, public funding of science is a relatively small fraction of the US federal budget, around 1 percent. Who would balk at such a meager outlay? Many are balking now. Not because science has failed, but because scientists have failed to explain what we do and why it matters. We have built entire careers on public support, yet with a few exceptions, we have neglected our responsibility to talk to the public. If we fail to engage, then less careful individuals may do so on our behalf.

Science has long had its cultural adversaries. From those who objected to the pronouncements of science on religious grounds (think Galileo) to relativists who view science as "just another way of thinking," there is a common misconception of what science *is*. To those outside of science, it may feel like an arcane world full of math, complicated theories, strange equipment, people in lab coats, and impenetrable jargon. In reality, what we call *science* is just a more disciplined version of what we would normally call *learning*.

Although I worry about the future of universities and scientific research, I want to approach the conclusion of this narrative with slightly more positivity. Science is in dark times, it's true, but I see myself as an exemplar of what can happen in the best of circumstances. I entered the academic field as a nerdy kid from a poor family who was completely unsure of myself. Obsessed with computers, science fiction, and obscure English authors, and excluded from the so-called heteronormative world of sports and other boy stuff, I was an outsider—a square peg constantly bumping up against round holes. What I longed for was a place where I could fit in. A place where I could not only exist but thrive. I found that place in university life, where I met other people who were just like me: fellow nerds who shared my interests, yes, and also a deep desire to see how things worked. In this sense, universities provide a welcoming environment for people who feel like they have no other place in the world.

I hope that this book can be, in whatever way stories can provide roadmaps, a manifesto for nerdy kids who are just like I was. The sciences can offer a place of discovery, and they can also be a landing pad. It's my hope that the person I am now can act as a role model for every lonely nerd out there who wonders whether they will ever make it out of elementary, middle, or high school unscathed. I'm here to announce that there is, in fact, hope, that the ending is bright, that there are so many possibilities at the end of the story.

And yes, I still believe that higher learning and the enterprise of scientific research require a degree of overhaul. I want more accountability among university administrators, perhaps by incentivizing them to continue their teaching and scholarship to some modest extent. Greater connection to the faculty and students would improve educational innovation and morale.

I want more transparency and just plain honesty: The system produces far more PhDs in biology, chemistry, and physics than can be employed in industrial research positions. Thus, those who do find a job are undervalued relative to their skills. Positions and salaries for professional PhD researchers—experts in experimental design, execution in the lab, and data analysis—should be expanded. The pace of innovation will accelerate if we stop insisting that everybody doing bench work must be a "trainee" (and thus be underpaid).

I want to empower the conscientious mentors, perhaps by increasing the proportion of NSF and NIH funding awarded directly to students and postdocs so that they can vote with their feet and work with faculty who are dedicated not only to scientific innovation but also to the personal and professional flourishing of the personnel who do the work.

I want students to be better protected and better cared for so that they are prepared to enter the world as their whole selves, not as lab rats. That's what we owe them, as educators and as people.

www.ingramcontent.com/pod-product-compliance
Lightning Source LLC
Chambersburg PA
CBHW020544160726
47991CB00002B/574